K



# Guidelines for drinking-water quality

SECOND EDITION

## *Volume 3*
## *Surveillance and control of community supplies*



World Health Organization
Geneva
1997

WHO Library Cataloguing in Publication Data

Guidelines for drinking-water quality.—2nd ed.

Contents: v. 3. Surveillance and control of community supplies
1. Drinking water – standards 2. Data collection – methods 3. Guidelines 4. Developing countries I. Title: Surveillance and control of community supplies
ISBN 92 4 154503 8 (v. 3) (NLM Classification: WA 657)





TYPESET IN HONG KONG
PRINTED IN MALTA
96/11027—Best-set/Interprint—8000

# Contents

# Preface

The first edition of *Guidelines for drinking-water quality* was published by WHO in 1984–1985 and was intended to supersede earlier European and international standards. Volume 1 contained guideline values for various constituents of drinking-water and Volume 2 the criteria monographs prepared for each substance or contaminant on which the guideline values were based; Volume 3 was concerned with the monitoring of drinking-water quality in small communities, particularly those in rural areas.

During the International Drinking-Water Supply and Sanitation Decade (1981–1990) considerable experience was gained in the surveillance and improvement of small-community supplies, notably through a series of demonstration projects supported by WHO. This new edition of Volume 3 of *Guidelines for drinking-water quality* reflects the experience of these and many other projects concerned with improving the quality of water services undertaken during the Decade.

A number of important principles were established in the first edition of Volume 3 of the *Guidelines* and these continue to form an important part of the second edition. They include the distinct and complementary roles of the water supplier and the surveillance agency; the unique nature of the problems associated with monitoring small-community supplies (especially in developing countries); the central role of the microbiological monitoring of supplies of this type; and the importance of ensuring that surveillance leads to engineering improvements and other remedial measures. Experience gained during the Decade has highlighted the importance of other fundamental concepts which have been incorporated into this new edition, including the need to consider not only drinking-water quality, but also all aspects of water-supply services that influence health, and to address the problems of small periurban areas not covered by such services.

While conditions vary from country to country as a result of differences in economic, geographical, cultural and social conditions, the strategies and procedures described here should nevertheless be widely applicable. Thus it is hoped that this Volume, like the first edition, will prove useful to all those concerned with drinking-water supply to small communities: environmental health inspectors, sanitary technicians, laboratory personnel, water engineers, planners and all those in the health and water-supply sector with managerial responsibility for improving water-supply services to communities. For the purposes of this publication, the term "communities" applies not only to villages and small private water supplies in rural areas but also to other centres of population within, or in close proximity to, urban centres.

# Acknowledgements

The preparation of this volume was begun at a Review Meeting on Surveillance of Community Supplies, held in Harare, Zimbabwe, on 24–28 June 1991, when a detailed outline was agreed. The first draft of Volume 3 was reviewed at the Final Task Group Meeting on the Revision of the *WHO guidelines for drinking-water quality*, held in Geneva on 21–25 September 1992, and a revised draft was subsequently finalized at a Meeting on Technical Revision of Volume 3, held in Tirana, Albania, on 15–20 June 1993. The final version is the outcome of the work of a number of contributors and reviewers whose names are given in Annex 1; their assistance is greatly appreciated. The coordinator for Volume 3 of the *Guidelines* was J. Bartram, Manager, Water and Wastes, WHO European Centre for Environment and Health, Rome, Italy, formerly of the Robens Institute of Health and Safety, University of Surrey, Guildford, England.

The first edition of Volume 3 of the *Guidelines* provided the basis for a number of pilot projects and country programmes in Central and South America, Africa, various parts of Asia and in the Pacific region, funded jointly by the United Nations Environment Programme (UNEP) and the United Kingdom Overseas Development Administration (ODA). Regional and national training courses were conducted, which were also supported by the Danish International Development Agency (DANIDA) and which allowed for the review and evaluation of the approaches proposed in the *Guidelines.* The experience gained in the projects in Indonesia, Peru, and Zambia was evaluated and published (Lloyd B, Helmer R. *Surveillance of drinking water quality in rural areas.* Harlow, Longman Scientific and Technical, 1991), and provided the basis for much of the revised methodology in the second edition, including an intensified sanitary-inspection process and a new hazard-analysis scheme.

The revision of Volume 3 of the *Guidelines* was made possible through a grant provided by ODA to the Robens Institute of Health and Safety, University of Surrey, Guildford, England. Financial support for the review meetings was provided by DANIDA.

# Acronyms and abbreviations used in the text

| | |
|---|---|
| CFU | colony-forming units |
| DPD | diethyl-*p*-phenylenediamine |
| ESA | external support agencies |
| HTH | high-test hypochlorite |
| ISO | International Organization for Standardization |
| JTU | Jackson turbidity unit |
| MF | membrane filtration |
| MPN | most probable number |
| MSD | minimum safe distance |
| MT | multiple tube |
| NA | not applicable |
| NGO | nongovernmental organization |
| NTU | nephelometric turbidity unit |
| PA | presence–absence test |
| TCU | true colour unit |
| UNCED | United Nations Conference on Environment and Development |
| WHO | World Health Organization |

# 1. Introduction

## 1.1 Scope and purpose

This volume of *Guidelines for drinking-water quality* describes the methods employed in the surveillance of drinking-water quality in the light of the special problems of small-community supplies, particularly those of developing countries, and outlines the strategies necessary to ensure that surveillance is effective. It is also concerned with the linkage between surveillance and remedial action and with the form that remedial action should take.

The structure of this volume reflects the key stages in the development of surveillance, as summarized in Fig. 1.1. Thus Chapter 2 covers planning, and subsequent chapters deal with the procedures used in the collection of information—sanitary inspection and community surveys (Chapter 3), and the analysis of water quality (Chapter 4). Chapter 5 considers the analysis and interpretation of the information gathered and its use in improving water-supply services. The final three chapters cover strategies for improvement—technical interventions (Chapter 6), hygiene education (Chapter 7) and legislation and regulation (Chapter 8).

## 1.2 Community water supplies

The precise definition of a "community water supply" will vary. While a definition based on population size or the type of supply may be appropriate under many conditions, it is often administration and management that set community supplies apart, and this is especially true in developing countries. The increased involvement of ordinary, often untrained and sometimes unpaid, community members in the administration and operation of water-supply systems is characteristic of small communities; this provides a ready distinction between community water supplies and the supply systems of major towns and cities. However, water supplies in periurban areas—the communities surrounding major towns and cities—are often organizationally similar to those of rural communities; these may also be classified as "community water supplies" and are therefore included in this volume.

While the safe quality of water supplied to communities is an important consideration in the protection of human health and well-being, it is not the only factor that affects consumers. *Access to water* is of paramount concern and other

***Fig. 1.1 Key stages in the development of water-supply surveillance and strategies for improvement***

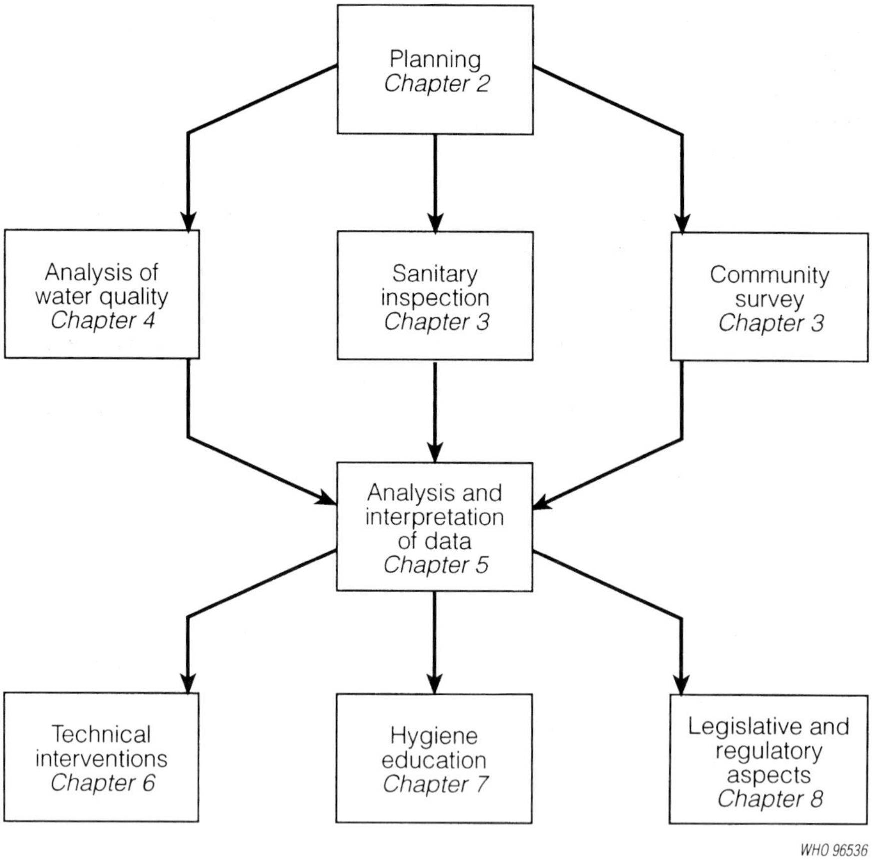


factors, such as the population served, the reliability of the supply and the cost to the consumer, must therefore be taken into account. At the United Nations conference at Mar del Plata in 1977, which launched the International Drinking-Water Supply and Sanitation Decade, this philosophy was adopted unambiguously: "*all peoples, whatever their stage of development and social and economic condition, have the right to have access to drinking-water in quantities and of a quality equal to their basic needs.*"

Access to water may be restricted in several ways, e.g. by prohibitive charges, daily or seasonal fluctuations in availability or lack of supplies to remote areas, and many countries face problems of this sort. In some parts of the world where water is scarce and has to be transported over long distances by road or on foot, the cost of drinking-water may absorb a significant proportion of the average daily income. Elsewhere, seasonal, geographical and hydrological factors may conspire to deny individual households or entire communities a continuous, reliable supply of drinking-water. During dry seasons, spring sources may dwindle, reservoirs may become exhausted and excessive demands by one group of people may limit supplies to their neighbours. Such problems are not confined to poorer countries; they are also experienced with increasing frequency in industrialized countries where management of demand has failed or population growth has outpaced the development of water resources.

If the performance of a community water-supply system is to be properly evaluated, a number of factors must be considered. Some countries that have developed national strategies for the surveillance and quality control of water-supply systems have adopted *quantitative service indicators* for application at community, regional and national levels. These usually include:

| | |
|---|---|
| quality: | the proportion of samples or supplies that comply with guideline values for drinking-water quality and minimum criteria for treatment and source protection |
| coverage: | the percentage of the population that has a recognizable (usually public) water-supply system |
| quantity: | the average volume of water used by consumers for domestic purposes (expressed as litres per capita per day) |
| continuity: | the percentage of the time during which water is available (daily, weekly or seasonally) |
| cost: | the tariff paid by domestic consumers |

Together, these five service indicators provide the basis for setting targets for community water supplies. They serve as a quantitative guide to the comparative efficiency of water-supply agencies and provide consumers with an objective measure of the quality of the overall service and thus the degree of public health protection afforded.

## 1.3 Health implications

The provision of an adequate supply of safe water was one of the eight components of primary health care identified by the International Conference on Primary Health Care in Alma-Ata in 1978. The guidelines presented here are in full accord with the spirit of the Alma-Ata declaration on primary health care, which expanded the concept of health care to include broader notions of affordability, accessibility, self-reliance, intersectoral collaboration, community participation, sustainability and social justice.[1]

In most countries the principal risks to human health associated with the consumption of polluted water are microbiological in nature (although the importance of chemical contamination should not be underestimated). As indicated in Chapter 18 of "Agenda 21" of UNCED, "An estimated 80% of all diseases and over one-third of deaths in developing countries are caused by the

[1] *Alma-Ata 1978: primary health care.* Geneva, World Health Organization, 1978.

consumption of contaminated water and on average as much as one-tenth of each person's productive time is sacrificed to water-related diseases."

The risk of acquiring a waterborne infection increases with the level of contamination by pathogenic microorganisms. However, the relationship is not necessarily a simple one and depends very much on factors such as infectious dose and host susceptibility. Drinking-water is only one vehicle for disease transmission. Some agents may be transmitted primarily from person to person and, for bacteria capable of multiplication in food, foodborne transmission may be more important than transmission by drinking-water. Other agents, however, such as *Salmonella typhi*, *Vibrio cholerae*, *Giardia lamblia* and hepatitis A virus, are frequently transmitted via contaminated drinking-water and, where this is the case, improvements in drinking-water quality may result in substantial reductions in disease prevalence.

Because of this multiplicity of transmission routes, improvements in the quality and availability of water, excreta disposal, and hygiene in general are all important factors in reducing diarrhoeal morbidity and mortality.

Epidemiological investigations indicate that all aspects of the quality of water supply services influence health, as do hygiene behaviours and sanitation. Experience has shown that analysis of disease incidence (epidemiological surveillance) is not a useful tool for guiding even large-scale remedial programmes for community water supplies. It is expensive and yields data that are difficult to interpret.

In the same way that indicators of the quality of water-supply services have been found useful in guiding remedial action, indicators of hygiene practices should also be used. Such indicators should be based on simple, standardized observations, and used to guide hygiene education programmes and the selection of key messages regarding hygiene behaviours.

### 1.3.1 Water quality

Guideline values for drinking-water quality are given in Volume 1 of the *Guidelines for drinking-water quality*, which also explains how the values should be interpreted. The health criteria used in establishing these values are summarized in Volume 2. A drinking-water quality guideline value represents the concentration of a constituent that does not result in any significant health risk to the consumer over a lifetime of consumption. Drinking-water should be suitable for human consumption and for all usual domestic purposes. When a guideline value is exceeded, the cause should be investigated and corrective action taken. The amount by which, and for how long, any guideline value can be exceeded without endangering human health depends on the specific substance involved.

In drawing up national standards for drinking-water quality, it will be necessary to take into account various local, geographical, socioeconomic and cultural factors. As a result, national standards may differ appreciably from the guideline values.

There may be a need for *interim standards* to provide a medium-term goal as a step towards the achievement of guideline values in the longer term. There is no objection to such a stepwise approach provided that the relevant authorities in each country, especially the ministry of health or its equivalent, are consulted and approve it. There are dangers in leaving such matters entirely to the agencies responsible for water supply because of the conflict of interests that may arise.

While supplies that fail to meet ideal criteria should be neither condoned nor ignored, interim standards permit resources to be directed first towards those communities with the greatest problems. They provide incentives to upgrade rather than blame for failure; this is particularly important in countries subject to severe economic constraints. The use of categories of bacteriological contamination of small-community supplies is useful in this context and is discussed in greater detail in Chapter 5.

In some countries, health authorities have adopted interim standards for intractable natural contaminants such as fluoride, pending the development of appropriate treatments for their removal from community supplies.

No attempt is made here to establish guideline values for service indicators other than drinking-water quality, such as those for the coverage, continuity, and cost of community water supplies. It is for national authorities to establish medium- and long-term targets for such factors. This should be done on a multisectoral basis, since the setting of these targets will have a number of social and economic implications. Nevertheless, because of the importance to public health of adequate access to safe water, the adoption of standards in this area is strongly recommended.

### Microbiological aspects

Ideally, drinking-water should not contain any microorganisms known to be pathogenic—capable of causing disease—or any bacteria indicative of faecal pollution. To ensure that a drinking-water supply satisfies these guidelines, samples should be examined regularly. The detection of *Escherichia coli* provides definite evidence of faecal pollution; in practice, the detection of thermotolerant (faecal) coliform bacteria is an acceptable alternative.

Guideline values for bacteriologically safe supplies of drinking-water are provided in Volume 1 of the *Guidelines.* Although developed for large water-supply systems, the values for treated and untreated water supplies are also applicable to community supplies and are therefore reproduced in Table 1.1. Background information on the significance and choice of indicator organisms, as well as the selection of analytical methods, is given in Chapter 4.

A complementary strategy for securing the microbiological safety of drinking-water supplies has also been advocated by WHO and a number of other agencies, based on the minimum treatment for certain types of water. This helps to ensure the elimination of faecal pathogens by specifying the conditions to be observed and treatments to be applied at the water-treatment plant. For example,

***Table 1.1 Guideline values for bacteriological quality*[a]**

| Organisms | Guideline value |
|---|---|
| **All water intended for drinking** | |
| *E. coli* or thermotolerant coliform bacteria[b,c] | Must not be detectable in any 100-ml sample |
| **Treated water entering the distribution system** | |
| *E. coli* or thermotolerant coliform bacteria[b] | Must not be detectable in any 100-ml sample |
| Total coliform bacteria | Must not be detectable in any 100-ml sample |
| **Treated water in the distribution system** | |
| *E. coli* or thermotolerant coliform bacteria[b] | Must not be detectable in any 100-ml sample |
| Total coliform bacteria | Must not be detectable in any 100-ml sample. In the case of large supplies, where sufficient samples are examined, must not be present in 95% of samples taken throughout any 12-month period |

[a] Immediate investigative action must be taken if either *E. coli* or total coliform bacteria are detected. The minimum action in the case of total coliform bacteria is repeat sampling; if these bacteria are detected in the repeat sample, the cause must be determined by immediate further investigation.

[b] Although *E. coli* is the more precise indicator of faecal pollution, the count of thermotolerant coliform bacteria is an acceptable alternative. If necessary, proper confirmatory tests must be carried out. Total coliform bacteria are not acceptable indicators of the sanitary quality of rural water supplies, particularly in tropical areas where many bacteria of no sanitary significance occur in almost all untreated supplies.

[c] It is recognized that, in the great majority of rural water supplies in developing countries, faecal contamination is widespread. Under these conditions, the national surveillance agency should set medium-term targets for the progressive improvement of water supplies.

cysts of protozoa such as *Giardia* and *Cryptosporidium* are efficiently removed by means of slow sand filters. Similarly, enteric viruses are inactivated by the maintenance of a disinfectant residual of at least 0.5 mg/litre free chlorine for a minimum of 30 minutes in waters with a turbidity of less than 1 NTU and a pH of less than 8.0. Another aspect of the minimum treatment approach is the protection of sources and catchments in order to minimize both contamination and the sophistication of the treatment processes needed to ensure potability.

It is not easy to provide generally applicable guidelines for other biological hazards, particularly parasitic protozoa and helminths. The application of any proposed guidelines and procedures must be governed by epidemiological considerations in at least two respects:

- Many parasites have a complex geographical distribution and it may be unnecessary to take precautions against those that do not occur locally.

- The majority of waterborne parasites are also transmissible by other routes, such as food and direct faecal–oral spread, and these routes should also be considered in the formulation of strategies for control.

Species of protozoa known to have been transmitted by the ingestion of contaminated drinking-water include *Entamoeba histolytica* (which causes amoebiasis), *Giardia* spp., and *Cryptosporidium.* These organisms can be introduced into a water supply through human or, in some instances, animal faecal contamination. Coliform organisms do not appear to be a good indicator of *Giardia* or *E. histolytica* in drinking-water: enteroviruses and protozoa are more resistant to disinfection than *E. coli*, so that absence of *E. coli* will not necessarily indicate freedom from these organisms.

The infective stages of many helminths such as parasitic roundworms and flatworms can be transmitted to humans through drinking-water. A single mature larva or fertilized egg can cause infection, and such infective stages should be absent from drinking-water. However, the water route is relatively unimportant except in the case of *Dracunculus medinensis* (the guinea worm), which is encountered mainly in unpiped water supplies. While there are methods for detecting this parasite, they are unsuitable for routine monitoring.

### Disinfection

Terminal disinfection is essential for surface waters after treatment and for protected groundwater sources when *E. coli* or thermotolerant (faecal) coliforms are detected. Chlorine in one form or another is the most commonly used disinfectant worldwide.

For terminal chlorination, there should be a free chlorine residual of at least 0.5 mg/litre after a minimum contact time of 30 minutes at a pH of less than 8.0, as for inactivation of enteric viruses. When chlorine is used as a disinfectant in a piped distribution system, it is desirable to maintain a free chlorine residual of 0.2–0.5 mg/litre throughout, to reduce the risk of microbial regrowth and the health risk of recontamination. In emergencies, e.g. in refugee camps, during outbreaks of potentially waterborne disease, or when faecal contamination of a water supply is detected, the concentration of free chlorine should be increased to greater than 0.5 mg/litre throughout the system.

High levels of turbidity can protect microorganisms from the effects of disinfection, stimulate the growth of bacteria, and give rise to a significant chlorine demand. Effective disinfection requires that turbidity is less than 5 NTU; ideally, median turbidity should be below 1 NTU.

Chlorine can be easily monitored and controlled as a drinking-water disinfectant, and regular, frequent monitoring is recommended wherever chlorination is practised. Chlorine determination is described in section 6.6.11. The health-based guideline value for free chlorine in water supplied to the public is

5 mg/litre. However, concentrations that are detectable by consumers and may provoke rejection may be much lower than this (typically 0.6–1 mg/litre); an upper limit should therefore be established based on local experience.

Disinfection is of unquestionable importance in the supply of safe water for drinking purposes. The destruction of microbial pathogens is essential and very commonly involves the use of reactive chemical agents such as chlorine. The use of chemical disinfectants usually results in the formation of chemical by-products, some of which are potentially hazardous, but the risks to health posed by these by-products are extremely small in comparison with those associated with inadequate disinfection. It is important that disinfection should not be compromised by attempts to control such by-products.

### *Chemical aspects*

In rural areas of developing countries, the great majority of health-related water-quality problems are the result of bacteriological or other biological contamination. Nevertheless, a significant number of very serious problems may occur as a result of the chemical contamination of water resources.

Some potentially chronic effects may occur in rural areas where overuse of agrochemicals leads to significant levels of pesticides in water sources. The presence of nitrate and nitrite in water may result from the excessive application of fertilizers or from leaching of wastewater or other organic wastes into surface water and groundwater. Although effects may be difficult to detect in human populations, such contaminants may pose a risk to health.

In areas with aggressive or acidic waters, the use of lead pipes and fittings or solder can result in elevated lead levels in drinking-water, which may, after long-term exposure, affect the mental development of children. Exposure to high levels of naturally occurring fluoride can lead to mottling of teeth and (in severe cases) skeletal fluorosis and crippling. Similarly, arsenic may occur naturally, and long-term exposure via drinking-water may result in a risk to health.

More acute health effects of chemical contamination of small-community supplies include methaemoglobinaemia in infants due to high levels of nitrate, and toxicosis due to accidental and other discharges of solvents and heavy metals from mining activities.

In order to establish whether or not this type of problem exists, a selected number of physicochemical parameters may have to be measured. However, it may be both very costly and physically impractical to cover a large number of parameters, particularly in the case of rural water supplies in developing countries.

If certain chemical contaminants are of special local significance, the levels should be measured and the results evaluated in the light of the guideline values and other recommendations made in Volume 1. It should also be noted that some health effects may occur as a result of specific chemical deficiencies in the diet, of which water forms a part. Important examples are ophthalmic goitre caused by iodine deficiency and dental caries resulting from low fluoride intake. No attempt has been made in these guidelines to define a minimum desirable concentration of such substances in drinking-water.

### Physical and aesthetic aspects

The chemical and physical quality of water may affect its acceptability to consumers. Turbidity, colour, taste, and odour, whether of natural or other origin, affect consumer perceptions and behaviour. In extreme cases, consumers may avoid aesthetically unacceptable but otherwise safe supplies in favour of more pleasant but less wholesome sources of drinking-water.

Although guidelines for drinking-water quality are based on the best available public health advice, there is no guarantee that consumers will be satisfied or dissatisfied by water supplies that meet or fail to meet those guidelines. It is therefore wise to be aware of consumer perceptions and to take into account both health-related guidelines and aesthetic criteria when assessing drinking-water supplies.

- *Turbidity* in excess of 5 NTU (5 JTU) may be noticeable and consequently objectionable to consumers.
- *Colour* in drinking-water may be due to the presence of organic matter such as humic substances, metals such as iron and manganese, or highly coloured industrial wastes. Experience has shown that consumers may turn to alternative, perhaps unsafe, sources, when their water displays aesthetically displeasing levels of colour, typically exceeding 15 TCU. Drinking-water should ideally be colourless.
- *Odour* in water is due mainly to the presence of organic substances. Some odours are indicative of increased biological activity, while others may originate from industrial pollution. Sanitary surveys should include investigations of sources of odour when odour problems are identified.

The combined perception of substances detected by the senses of taste and smell is often called "taste". "Taste" problems in drinking-water supplies are often the largest single cause of consumer complaints. Changes in the normal taste of a public water supply may signal changes in the quality of the raw water source or deficiencies in the treatment process.

Water should be free of tastes and odours that would be objectionable to the majority of consumers.

### Critical parameters of drinking-water quality in community supplies

The principal risks to human health associated with community water supplies are microbiological, and it has been traditional to rely on relatively few water-quality tests to establish the safety of supplies. Some agencies refer to this strategy as "*minimum monitoring*", while others use the term "*critical-parameter testing*".

The approach is based on the assumption that health authorities will be aware of other specific sources of risk in each region, such as chemical contamination, and will include these in the monitoring scheme. It is much more effective to test for a narrow range of key parameters as frequently as possible (*in conjunction* with a sanitary inspection) than to conduct comprehensive but lengthy and largely irrelevant analyses less frequently.

The parameters recommended for the minimum monitoring of community supplies are those that best establish the hygienic state of the water and thus the risk (if any) of waterborne infection. The critical parameters of water quality are thus:

- *E. coli*; thermotolerant (faecal) coliforms are accepted as suitable substitutes;
- chlorine residual (if chlorination is practised).

These should be supplemented, where appropriate, by:

- pH (if chlorination is practised);
- turbidity (if *any* treatment is effected).

The value and application of these tests are described in greater detail in Chapter 4. However, an advantage worth noting here is that these critical parameters may be measured *on site* using relatively unsophisticated testing equipment. On-site testing is essential for the determination of turbidity and chlorine residual, which change rapidly during transport and storage; it is also important for the other parameters where laboratory support is lacking or where transportation problems would render conventional sampling and analysis difficult or impossible.

Water suppliers need to carry out a wider range of analyses relevant to the operation and maintenance of water-treatment and distribution systems, in addition to the health-related parameters laid down in national water-quality standards. Analyses should also embrace the concept of acceptability: Volume 1 indicates that water supplied for drinking purposes should be inoffensive to consumers. Consumers may resort to a more palatable, but possibly unsafe, source if water is considered unacceptable; acceptability is therefore also considered a critical parameter. It may be assessed by observation (taste, colour, odour, visible turbidity) and requires no laboratory determinations.

Other health-related parameters of local significance should also be measured. It may sometimes be useful to include total coliforms in the bacteriological analysis, e.g. if chlorination is practised and there is an extensive distribution network.

### Other important analyses

When supply sources are being investigated for the first time or when new sources are being developed, it is prudent to undertake a wide range of analyses in order to establish the overall safety and wholesomeness of the water.

It is essential that all water-quality factors are taken fully into account when technologies for abstraction and treatment of new resources are selected. Seasonal variations in the turbidity of raw surface waters can be very great, and allowance must be made for this; treatment plants should be designed for worst-case conditions rather than for average water quality, otherwise filters may rapidly become blocked or sedimentation tanks overloaded. The chemical aggressiveness of some groundwaters may affect the integrity of borehole casings and pumps, leading to unacceptably high levels of iron in the supply, eventual breakdown, and expensive repair work. Both the quality and availability of water may then be reduced and public health endangered.

In most water sources, especially groundwaters, the majority of chemical parameters vary relatively little with time. Thus, for routine assessments, it is advisable to investigate those parameters most closely related to health risk and/or most liable to change over short periods.

### 1.3.2 Water-washed diseases

A reliable, safe water supply plays an important role in disease prevention, especially by facilitating personal, domestic, and food hygiene. The diseases most affected by the provision of adequate quantities of water for hygienic purposes are referred to as *water-washed.* They may be divided into the following three groups:

- Diseases transmitted by the faecal–oral route, such as hepatitis A, bacillary dysentery, and many diarrhoeal diseases; these are transmitted by water and also by other means, such as food or hands. Improved hygiene therefore contributes to their control.
- Infections of the skin and eyes, such as trachoma, skin infections, and fungal skin diseases. The prevalence of these diseases is related to poor hygiene.
- Infections carried by lice or mites, such as scabies (mites), and louse-borne epidemic typhus (caused by *Rickettsia prowazeki* and transmitted largely by body lice). Good personal hygiene can assist in control.

Provision of water for domestic purposes in adequate quantities and quality will contribute to reducing the incidence of diseases transmitted by the faecal–oral route and other transmissible diseases.

## 1.4 Objectives of surveillance and quality control

Surveillance is an investigative activity undertaken to identify and evaluate factors associated with drinking-water which could pose a risk to health. Surveillance contributes to the protection of public health by promoting improvement of the quality, quantity, coverage, cost, and continuity of water supplies. It is also both preventive—detecting risks so that action may be taken before public health problems occur—and remedial—identifying the sources of outbreaks of waterborne disease so that corrective action may be taken promptly.

Surveillance requires a systematic programme of surveys that combine analysis, sanitary inspection, and institutional and community aspects. Sanitary inspection should cover the whole of the water-supply system including sources, conduction lines, treatment plants, storage reservoirs, and distribution systems.

Surveillance is indispensable for the development of rational strategies for the improvement of the quality of water-supply services.

Quality control is designed to ensure that water services meet agreed national standards and institutional targets.

Water suppliers are responsible at all times for the quality and safety of the water that they produce, and they achieve this by a combination of good operating practice and preventive maintenance, supported by quality control. Water-quality control is the responsibility of the water supplier and involves the establishment of safeguards in the production and distribution of drinking-water as well as the routine testing of water quality to ensure compliance with national standards.

Quality control is distinguished from surveillance on the basis of institutional responsibilities and the frequency of the monitoring activities conducted. The surveillance agency is responsible for an independent (external) and periodic audit of all aspects of safety, whereas the water supplier is responsible at all times for regular quality control, and for monitoring and ensuring good operating practice.

## 1.5 Organizational structure

Organizational arrangements for the improvement of water-supply services should take into account the vital and complementary roles of the agency responsible for surveillance and of the water supplier. The two functions outlined in section 1.4, i.e. surveillance and quality control, are best performed by separate and independent entities because of the conflict of interests that arises when the two are combined. Nevertheless, because the two are essentially complementary, the monitoring of water-supply services should involve both the surveillance agency and the supplier.

Important aspects of a surveillance programme include the following:

- The surveillance agency should have sole responsibility within the health authority for providing surveillance services to protect the public from waterborne diseases and other hazards associated with the water supply.
- Water-supply surveillance should be integrated with other environmental health measures, especially sanitation.
- Surveillance requires specialized knowledge, and the agency should thus include personnel specially trained in sanitary engineering, community health, epidemiology, chemistry, biology, etc. Additional support should

be provided by the medical profession, particularly during an outbreak of enteric disease.
- Health authorities should have centralized laboratories and other services needed for programmes of water-supply surveillance.
- Periodic reports to the government regarding the public health status of the country's water supplies should be produced.

In countries where urban water suppliers have established effective quality control, the surveillance agency may choose to place greater emphasis on the problems of the less well served populations. Such populations are specifically addressed in this publication, and include both rural communities and urban fringe areas.

### 1.5.1 The surveillance agency

In most countries the agency responsible for the surveillance of drinking-water supply services is the ministry of health (or public health) and its regional or departmental offices. In some countries there is an environmental protection agency; in others, the environmental health departments of local government may have some responsibility. The surveillance agency should preferably be an established institution designated by national legislation, should be represented at national level, and should operate at central, provincial (departmental/regional), and local (district) levels. Its responsibilities should encompass the monitoring of compliance with supply service standards (including quality, coverage, quantity, continuity, and cost) by water suppliers, approving sources of drinking-water, and surveying the provision of drinking-water to the population as a whole.

Surveillance is concerned with *all* water used for domestic purposes by the population, whether supplied by a formal water-supply agency or collected from individual sources or supplies. The agency's area of responsibility should therefore embrace all sources of water used, or intended for use, for human consumption. Nevertheless, in many developing countries, especially where there are many sources that may each supply a small population, such a goal may be difficult and expensive to achieve. Priority should therefore be given to systems that provide water to larger populations and those suspected of causing a substantial risk to human health, and to the identification of the most common risks and shortcomings in the supplies.

### 1.5.2 Quality control and the role of the water supplier

What is said above does not exclude water-supply and construction agencies from involvement in surveillance; in fact, it is vital that they should be involved. While it is the responsibility of the surveillance agency to generate and summarize surveillance data and to promote improvements, it is the water-supply sector that will carry out many of the actions designed to improve supplies. In addition, supply and construction agencies are responsible for quality control of the service they provide. However, there may not always be a clear division of responsibilities between the health and water-supply sectors. In some cases, the range of professional, governmental, nongovernmental, and private institutions may be wider and more complex than that discussed above. Whatever the existing framework, it is important that clear strategies and structures are developed for implementing surveillance and quality control, collating and summarizing data, reporting and disseminating the findings, and taking remedial action. Similarly, clear lines of accountability and communication are essential.

The organizational arrangements for carrying out surveillance and quality-control activities may be modified as programmes move from the pilot stage to regional and then national implementation. It is important that basic local, regional, and national frameworks should be in place from the outset in order to avoid subsequent confusion, but they may well be refined and improved in the light of experience during the implementation of activities. It is preferable to develop and build on existing frameworks than to impose radical changes immediately before or during a programme.

## 1.6 Community participation

Community participation is an essential component of the surveillance framework. As primary beneficiaries of improved water supplies, community members have a right to take part in decision-making about their own future. They represent a resource that can be drawn upon for local knowledge, experience, financial support, and labour. They are the people who are most likely to notice problems in the water supply first and can therefore take immediate remedial action. Establishing a genuine partnership with the community creates a climate of trust and understanding, which itself generates interest and enthusiasm. This provides a good foundation for other educational activities such as the promotion of latrines and of good hygiene practices.

The community's role in the planning and implementation of surveillance can valuably include the following:

- — assisting in the establishment of procedures for surveillance;
- — assisting in data collection;
- — assisting field workers in water sampling;
- — monitoring water quantity and quality and regularly reporting findings to surveillance staff;
- — ensuring appropriate use of water supplies;
- — setting priorities for remedial action, including improvement of water supplies, sanitation, and hygiene;
- — undertaking simple maintenance and repairs;
- — referral of problems that require special attention.

In involving the community in surveillance it is important to:

— provide an effective method, easily used by volunteers, to identify sanitary hazards associated with the water supply;
— provide training to community members in undertaking sanitary surveys and remedial action, and provide long-term support for such training in order to ensure sustainability.

## 1.7 Role of surveillance in improvement of water supplies

For water-supply surveillance to lead to improved drinking-water supply services it is vital that the mechanisms for promoting improvement are recognized and used.

A checklist of mechanisms for water-supply improvement based on the output of surveillance is given in Table 1.2. Similar concepts can be applied to

***Table 1.2 Mechanisms for the improvement of water-supply services based on the results of water-supply surveillance***

- **Establishing national priorities**
  When the commonest problems and shortcomings in water-supply systems have been identified, national strategies can be formulated for improvements and remedial measures; these might include changes in training (of managers, administrators, engineers, or field staff), rolling programmes for rehabilitation or improvement, or changes in funding strategies to target specific needs.
- **Establishing regional priorities**
  Regional offices of water-supply agencies can decide which communities to work in and which remedial activities are priorities; public health criteria should be considered when priorities are set.
- **Establishing hygiene education**
  Not all of the problems revealed by surveillance are technical in nature, and not all are solved by supply and construction agencies; surveillance also looks at problems involving private supplies, water collection and transport, and household treatment and storage. The solutions to many of these problems are likely to require educational and promotional activities coordinated by the health agency.
- **Enforcement of standards**
  Many countries have laws and standards related to public water supply. The information generated by surveillance can be used to assess compliance with standards by supply agencies. Corrective action can be taken where necessary, but its feasibility must be considered, and enforcement of standards should be linked to strategies for progressive improvement.
- **Ensuring community operation and maintenance**
  Support should be provided by a designated authority to enable community members to be trained so that they are able to assume responsibility for the operation and maintenance of their water supplies.

water-quality control programmes, but it is then likely that greater emphasis will be placed on the setting of investment priorities at regional and national levels than on hygiene education and enforcement. Each of the mechanisms is discussed in greater detail in subsequent chapters.

Information alone does not lead to improvement. It is the effective management and use of the information generated by surveillance that makes possible the rational improvement of water supplies—where "rational" implies that available resources are used for maximum public health benefit.

# 2. Planning and implementation of surveillance

## 2.1 Legal and institutional basis

### 2.1.1 Laws, regulations, and standards

Effective programmes to control drinking-water quality depend ideally on the existence of adequate legislation, standards, and codes. One of the functions of the basic legislation is to define the functions, authority, and responsibilities of the water-supply agency and the surveillance agency. Standards and codes should specify the quality of the water to be supplied to the consumer, the practices to be followed in selecting and developing water sources and in treatment processes and distribution systems, and procedures for approving water systems in terms of water quality. The precise nature of the legislation in each country will depend on national, constitutional, and other considerations.

Experience has shown that the basic legislation should be limited to general principles and to specifying offences and penalties for its contravention. The authority to establish and revise drinking-water standards, codes of practice, and other technical regulations should be delegated to the appropriate government minister—preferably the minister of health—who is responsible for ensuring the quality of water supplies and the protection of public health. The authority to establish and enforce quality standards and regulations may be vested in a ministry other than that usually responsible for public and/or environmental health. Consideration should then be given to requiring that water-quality standards are promulgated only after approval by the public health or environmental health authority so as to ensure their conformity with health-protection principles.

Such legislation commonly makes provision for the establishment and amendment of drinking water quality standards and guidelines, as well as regulations for the development of drinking-water sources, and the production, maintenance, and distribution of safe drinking-water. It also generally establishes the legal functions and responsibilities of the water-supply agency, and states clearly that, as an organization that sells and/or supplies water to the consumer, this agency has a legal duty to supply safe and wholesome water that meets legally established water-quality standards. In addition, the agency is responsible for providing continuous and effective quality assurance and quality control of water supplies, including inspection, supervision, preventive maintenance, routine testing of water quality, and remedial actions as required.

The water-supply agency should be deemed responsible for the safety and quality of the water supply up to a defined point in the distribution system, generally the house connection or public standpost.

A country-wide or regional water-supply company or governmental organization often supplies drinking-water to a municipal water-supply agency or a local water-distribution company or group. As the "wholesaler", the primary supplier should be legally responsible for the water quality up to the point of connection to the pipelines of the local supplier; the organization that supplies the public directly then becomes the "retailer". In other words, each organization should carry legal responsibility for the quality of the water supply up to the point of delivery to the "customer".

Governments should also consider introducing legislation that would enable individuals or community organizations to take legal action to enforce water-quality standards and regulations. They should consider making legal provisions for water-supply agencies to initiate legal action to protect their water sources and distribution systems from sources of pollution. This is particularly important in areas where no effective government programme is in operation to control pollution.

The surveillance agency should be given the necessary powers to administer and enforce laws, regulations, standards, and codes concerned with water quality. It should also be able to delegate those powers to other specified agencies such as municipal councils, local health departments, regional authorities, nongovernmental (community) organizations, universities, and qualified, government-authorized private testing services.

Many countries lack basic legislation of this sort, and in others the existing legislation is seriously outdated. However, many interim measures to ensure drinking-water quality can be enforced under existing general health, food, and welfare legislation. Implementation of programmes to provide safe drinking-water should not be delayed because of a lack of appropriate legislation.

Even where legally binding guidelines or standards for drinking-water have yet to be promulgated it may be possible to encourage, and even enforce, the supply of safe drinking-water through educational efforts or commercial, contractual arrangements between customer and supplier based on civil law.

The application of water-supply legislation is considered in Chapter 8.

### 2.1.2 Institutional framework for water-quality surveillance

The main role of surveillance in the management of community water supplies is to assess the safety and acceptability of the water distributed to the public so that consumers are consistently and reliably protected from the health hazards of contaminated supplies. Surveillance therefore adds considerably to the value of water, especially for domestic use. It facilitates the recovery of its cost and increases its health benefits.

Water-quality surveillance requires an institutional framework that reflects its objectives and functions and gives key responsibilities to the relevant bodies—not just the agencies in charge of supplying water and promoting health but all institutions with relevant normative, developmental, educational, and control functions.

At the centre of this framework major responsibility for surveillance is shared between two agencies whose activities should be both mutually exclusive and complementary. The water-supply agency is responsible for the quality and safety of the water that it produces and distributes, while the surveillance agency has overall responsibility for ensuring that all drinking-water supplies under its jurisdiction are free from health hazards. Indirectly, however, health hazards related to the ingestion or other utilization of contaminated water from unprotected sources may be the fault of the water-supply agency if it has failed to fulfil its mandate, thus causing the public to use unsafe supplies.

The water-supply agency also differs from the surveillance agency in the sense that it carries out routine testing and monitoring of the quality of the water that it produces, while the public health protection agency conducts independent surveillance audits of water quality to determine whether the water-supply agency is fulfilling its responsibility.

The key basic principle in the implementation of a reliable programme of surveillance of drinking-water quality is that this two-tier system is absolutely necessary. It is imperative that the public health protection agency is adequately equipped to fulfil its regulation functions. If it is not, surveillance tasks can be subcontracted by the surveillance agency to a third party, such as a private company, at a cost that can be recovered, e.g. in the selling price of water. Monitoring by the water-supply agency of the quality of its own product, or that of an affiliated company, should never be considered as a satisfactory substitute for independent surveillance.

Another important principle is that the institutional and legal arrangements for water-quality surveillance should lend themselves to both decentralization and intersectoral cooperation. Like all water-related activities, water-quality promotion and control are fragmented horizontally between a large number of producers, users, and planning, financing, and monitoring agencies, as well as vertically between national and regional agencies with limited potential for decentralization, numerous local authorities with scarce resources, and a very large number of consumption points, especially in developing countries.

Intersectoral cooperation is required in all activities related to the promotion and surveillance of water quality, from the normative functions to the actual supply of water, the surveillance of water quality, and the implementation of preventive and remedial measures. At the normative stage, those agencies responsible for the protection of public health and for the supply of water should, in consultation with one another, agree on safe and feasible water-quality standards. To ensure that these standards are also acceptable to consumers, the communities served should always be involved, together with the major water users, including industrial or agricultural concerns that may compete for the same water source or public supply.

Other normative and regulatory functions belong to such ministries as those responsible for public works, housing, natural resources, or the environment, which are concerned with the design of water-supply and waste-disposal systems, equipment standards, plumbing codes and rules, water allocation, protection and conservation, and waste collection, treatment, and disposal. The economic planning unit (for resources allocation), the ministries and agencies in charge of internal affairs and local government (for community issues), and the ministry of finance (in relation to water tariffs) should be consulted on issues within their respective areas of competence. Private autonomous water suppliers should also be involved in drawing up standards if this is justified by their individual or collective size and importance; the national regulations, adjusted as necessary, should always be applicable to such water suppliers. Successful intersectoral coordination requires the involvement of agencies responsible for community development and hygiene education in all activities and at all levels; these agencies are usually more easily decentralized than the water authorities. Public health agencies are often closer to the community than those responsible for its water supply. At local level, they also interact with other sectors, e.g. education, and their combined action is essential to ensure active community involvement.

Public health surveillance teams operate at national, provincial, and district levels, as well as in cities or at rural health centres. However, public health laboratories may be available only in large cities, in which case the use of field kits for water-quality testing (see pp. 65–66) by mobile surveillance teams may help to bridge the gap between fixed laboratories and remote communities.

Where they are able to operate in remote areas with widely scattered populations, surveillance teams can also provide essential epidemiological information that can be used in planning, and information on major faults that is valuable in the organization of maintenance. Where water-quality surveillance teams cannot operate, nongovernmental organizations can help, and community volunteers can also be trained. In some countries, religious missions, aid agencies, and scientific institutes play important roles in water-quality surveillance.

## 2.2 Planning

### 2.2.1 General considerations

To be successful, water-supply surveillance and quality control must be well planned, and the definition of objectives is fundamental to any planning process. In addition to the main objectives of surveillance and quality control identified in section 1.4, there will be a number of complementary objectives. These will vary according to the conditions under which the activities are to be implemented and will most commonly encompass the activities to be undertaken during implementation. Examples include the following:

- — provision of equipment and training;
- — determination of trends in the quality of the drinking-water supply service with time as shown by specific indicators;
- — provision of information to public authorities for general public health protection purposes (i.e. information dissemination);
- — identification of sources of contamination;
- — investigation of piped distribution networks;
- — identification of remedial strategies;
- — assessment of the performance of water-treatment plants;
- — involvement of communities in the surveillance process.

Targets provide the link between objectives and the plan of work, and should be reviewed at regular intervals, perhaps annually. In developing a surveillance programme, early targets would typically include:

- — preparation of a comprehensive water-supply inventory;
- — development of preliminary standard methodologies (e.g. for analytical procedures, field work, and reporting);
- — establishment of regional laboratories capable of undertaking specified analyses;
- — training of staff responsible for water sample analysis at regional and local levels;
- — preliminary survey visits to a number of communities, and involving community members in surveys and briefings as a preparation for their role in community-based surveillance;
- — implementation targets such as coverage (number of communities visited);
- — analysis of the data produced and dissemination of the findings to each community, to the local and regional authorities, to the water-supply and health agencies at regional and national levels, and to a national institution responsible for planning and coordination;
- — community-based education in hygiene.

Surveillance should clearly not be limited to data collection. For example, if it is noted that there is a particular need to promote public involvement in questions of water supply or to undertake appropriate health education, it may be decided that particular emphasis should be placed on these activities. It is important to ensure that specific objectives and targets are not overambitious: they should be clearly defined and achievable within a sensible, defined time-scale.

Objectives should not be established in the capital city and imposed on those required to implement the programme nationally. They must be discussed and agreed at all levels following a period of genuine and broad-based consultation, starting at the community level. If people are committed to a common goal and a common set of objectives, many of the problems commonly encountered during implementation will be overcome simply and with good will.

### 2.2.2 Strategies

The community management of water-supply services was one of the basic principles laid down at the Global Consultation on Safe Water and Sanitation in New Delhi in September 1990, which marked the end of the International Drinking-Water Supply and Sanitation Decade. Application of this principle implies that decisions must be taken at the lowest appropriate level, after public consultation and with the involvement of users in the planning and implementation of water-supply projects. Government programmes should provide assistance and support to communities in managing their own water-quality control systems.

The implementation of water-supply programmes and the accompanying surveillance is a national responsibility. To a varying degree, responsibility for the operation of supply and surveillance systems should be delegated to all administrative levels, down to the community and the individual served. National authorities should therefore develop mechanisms for collaboration at all levels; this is particularly important if full advantage is to be taken of community-based approaches and self-reliance as tools for achieving sustainability. Women must be involved in all aspects of water-supply and surveillance systems, including planning, decision-making, implementation, and evaluation. In addition, broad-based educational programmes should be established, with particular emphasis on hygiene, local management, and risk reduction.

Where it represents a new activity for health or environmental-protection agencies, the implementation of surveillance activities should begin at the pilot level, progress to regional level, and then expand to national level. The principle of initial pilot-scale implementation is important and has been found to be widely applicable. Other procedures for progressive implementation also exist; thus it may sometimes be appropriate to begin with larger centres of population and work down to small-community supplies. In both cases, it is important for activities to be initiated on the pilot scale and to be subject to evaluation and improvement.

Any approach in which extension to the national level takes place too rapidly has a number of potential disadvantages. This is especially true where implementation at pilot and regional levels depends on a national authority. In these circumstances, extension to a national surveillance or quality-control programme may make sudden and severe demands on the human and financial resources of this body. Careful preparation in terms of training and resource provision is always required.

Quality-control activities should be initiated as each new supply system is constructed, and should be continued on a routine basis thereafter. There is thus no question of a staged implementation of these activities unless the quality-control function has never been initiated or has collapsed and requires rehabilitation. Only the progressive implementation of water-supply surveillance is considered here, since in many countries it may represent a new activity. However, much of the detail concerning inventories, the design of forms, training, and field work is also relevant to the quality-control activities of a water supplier.

The limited availability of resources (especially in developing countries) makes it advisable to start surveillance with a basic programme that develops in a planned manner. Activities in the early stages must generate enough useful data to demonstrate the value of surveillance. Thereafter, the objective should be to progress to more advanced surveillance as resources and conditions permit. Three distinct phases may be identified—initial, intermediate, and advanced. The activities associated with each phase are summarized in Table 2.1.

## 2.3 Implementation

Surveillance activities differ from country to country and region to region, between urban and rural communities, and according to the types of water supply. They should be adapted to local conditions and to the availability of local financial resources, personnel, infrastructure, and knowledge-base. Factors influencing surveillance activities include:

- the type and size of water-supply systems;
- the equipment, both existing and available;
- local employment practices, and the level of training of personnel;
- opportunities for community participation;
- geographical conditions (e.g. the accessibility of systems);
- climatological conditions (which may hamper activities during certain seasons);
- communication and transport infrastructure.

In practice, the sequence of activities in the development of surveillance is usually similar to that summarized in Fig. 2.1.

### 2.3.1 Inventories

Methods of providing drinking-water vary widely. They may include the use of piped supplies with or without treatment and with or without pumping (supplied via domestic connection or public standpipe), delivery by tanker truck or carriage by pack animals, or collection from groundwater sources (springs or wells) or surface sources (lakes, rivers, and streams). All members of the population receive water by some means, and it is important for the surveillance agency to build up a picture of the frequency of use of the different types of supply, especially as a preliminary step in the planning of a surveillance programme. There is little to be gained from undertaking the surveillance of piped water supplies alone if these are available to only a small proportion of the population. Although the supply agency should be responsible for the quality control of all its supplies, its water sources will only rarely include open dug wells and private supplies, which may be more highly contaminated. For these sources surveillance is of paramount importance.

## *Table 2.1 Activities to be undertaken in the initial, intermediate, and advanced phases of water-supply surveillance*

**Initial phase**

- Establish requirements for institutional development
- Provide training for staff involved in programme
- Start inventories of supply systems
- Undertake sample surveys to identify priority areas
- Develop methodologies suitable for the area
- Commence routine surveillance in priority areas
- Limit water-quality analysis to critical parameters and known problem substances
- Establish reporting, filing, and communications systems (paper-based, rather than computerized)
- Make improvements according to identified priorities
- Establish reporting to local suppliers, communities, and regional authorities
- Establish liaison with communities; identify community roles in surveillance and means of promoting community participation

**Intermediate phase**

- Train staff involved in programme
- Complete inventories of supply systems
- Establish and expand systematic routine surveillance
- Expand analytical capability (often by means of regional laboratories, national laboratories being largely responsible for analytical quality control and training of regional laboratory staff)
- Undertake surveys for chemical contaminants using wider range of analytical methods
- Evaluate all methodologies (sampling, analysis, etc.)
- Use draft standard methods (e.g. analytical methods, fieldwork procedures)
- Establish (and possibly computerize) database archive
- Identify common problems, improve activities to address them at regional and national levels
- Expand reporting to include interpretation at national level
- Draft or revise national standards and legislation
- Use legal enforcement where possible
- Involve communities routinely in surveillance implementation

**Advanced phase**

- Train staff involved in programme
- Establish routine surveillance for all health and acceptability parameters at defined frequencies
- Use full network of central, regional, and local laboratories (including analytical quality control)
- Use national standards and legislation
- Improve water services on the basis of national and local priorities, hygiene education, and enforcement of standards
- Disseminate data at all levels (local, regional, and national)
- Involve communities routinely in surveillance implementation

***Fig. 2.1 Sequence of activities in the development of surveillance***

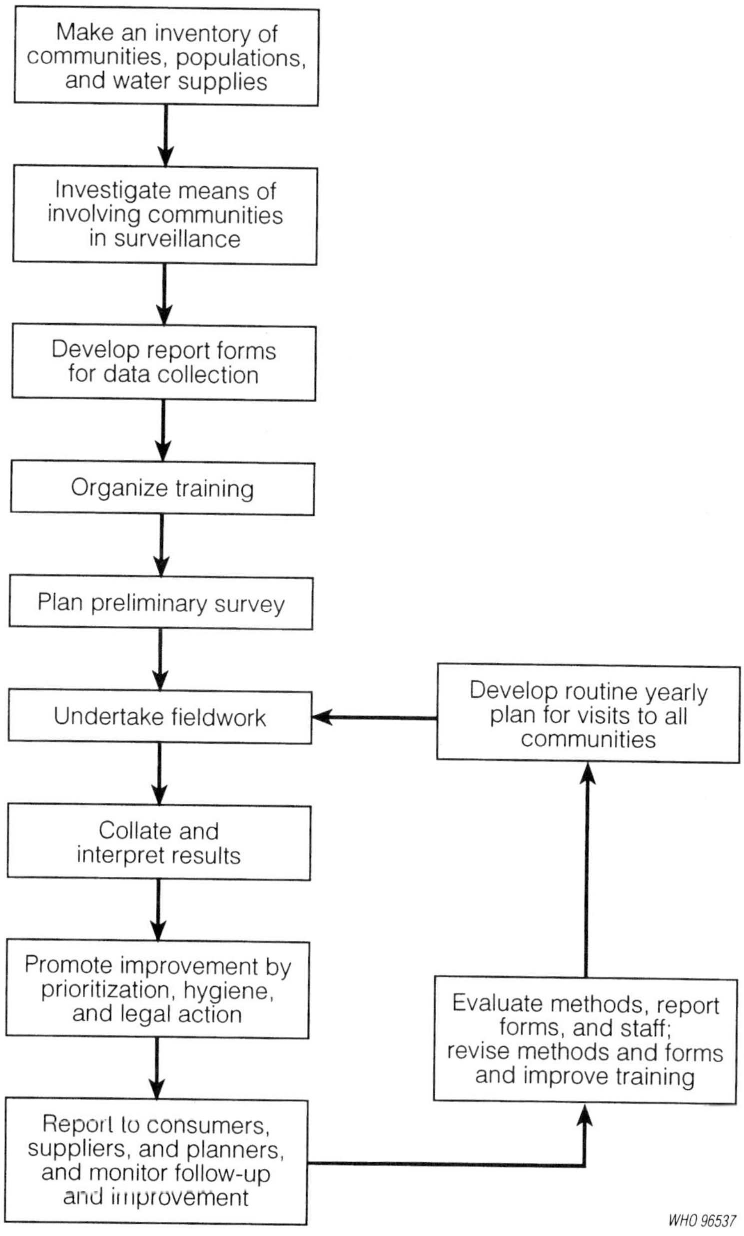

An inventory of all existing water-supply systems should be prepared by the surveillance agency, drawing on local community knowledge. It should include a register of communities, together with data on their total population; this information is often available through censuses. Descriptions of all known water-supply systems should be included, with details of physical components, administrative arrangements, and population served, supplemented by other relevant information (for example, on access and transport requirements).

Some of the information required for an initial inventory may be collected by means of a desk study but should be checked in the field through inspection and meetings with community groups and local leaders. This is especially important when information on water-supply systems is obtained from a central agency. It is then almost inevitable that the records will be incomplete, because systems may have been constructed by different agencies, possibly before records were kept, or by individuals or the communities themselves. A typical form for use in making an inventory of community water supplies is illustrated in Fig 2.2; it should be adapted according to local circumstances.

Where the initial inventory fails to show the means of supply to a significant proportion of the population, a survey should be undertaken to determine whether the information is incomplete and the means by which water is supplied to the remainder of the population.

In addition to a general overview of how the population obtains water for domestic purposes, the inventory provides a preliminary assessment of the workload of the surveillance agency and the field support required to involve the community in surveillance. This enables the cost of implementation to be estimated. It is important for such estimates to be realistic. Since it is unlikely that all existing supplies will have been identified, additional surveys may be necessary if, for example, open dug wells are found to be an important source of water and therefore to merit attention. It is also unlikely that estimates of travel time and transportation requirements will be accurate, and some allowance should be made for errors.

## 2.3.2 Designing forms

Simple-to-use community survey and sanitary inspection forms must be carefully developed. These should take the form of pictorial or written checklists that ensure standardized responses and assist the person doing the work to make a rapid assessment of water-supply service quality. Examples of report forms are given in Fig. 2.2 (community surveys) and Annex 2 (sanitary inspection); their design is described in detail in section 3.3. However, community water supplies vary widely, and it is important to evaluate the forms in the light of, and adapt them to suit, specific regional or national conditions. Where appropriate, separate forms should be designed for use by communities in assessing their own water supplies.

The sanitary survey report form may include details that also appear on the inventory, but since the information in the latter is not likely to change very much, it is more practical for the two to be separate. The sanitary survey report form should contain sanitary inspection details and therefore at least a checklist of the components of the system, including those for which the risk of contamination is greatest. The form should also include assessments of water-supply service indicators other than water quality, namely cost, coverage, continuity, and quantity.

In countries where there are many different types of supply, several different sanitary survey report forms may be necessary; a standard form may otherwise run to several pages.

The design, evaluation, and revision of sanitary survey report forms is important in the development of a surveillance programme. Only essential information should be collected, so that field staff are not burdened with collecting superfluous data. The order in which questions are arranged should coincide with that in

***Fig. 2.2 Typical form for making an inventory of community water supplies***

Date of vist . . . . . . . . .  Name of community: . . . . . . . . . . . . . . . . . . . . . . .

Agency responsible for supply . . . . . . . . . . . . . . . . . . . . . .
Community representative responsible for supply . . . . . . . . . . . . . . . . . . . . . .

Total population: . . . . . . . . . . . . . . . .
Population served by:
— house connections . . . . . . . . . .
— standposts . . . . . . . . . .
— protected springs/wells . . . . . . . . . .
— other . . . . . . . . . .

Distance to monitoring base: . . . . . . . km
Traval time from monitoring base: . . . . . . hours, by . . . . . . . . . . . . . . . . . . . (means of travel)

For piped systems:
— source type:
— treatment components:
  — infiltration galleries Y/N
  — surface-water intake Y/N
  — sedimentation Y/N
  — prefiltration Y/N
  — slow sand filtration Y/N
  — aeration Y/N
  — disinfection Y/N
— number of reservoirs . . . . . .
— number of standposts . . . . . .
— number of household connections . . . . . .

Health post/centre Y/N
School Y/N

which work is to be undertaken. Where there is to be on-the-spot reporting, the form should incorporate or be accompanied by an appropriate extra section for reporting the community responsible for, or the caretaker of, the supply. Clearly worded questions that will yield yes/no answers should be used wherever possible: standardized answers permit statistical analysis, which minimizes subjectivity in reporting and maximizes the usefulness of findings.

### 2.3.3 Training

The quality of the information produced by a surveillance programme will depend largely on that of the work undertaken by the staff responsible for liaison with communities, filling in the sanitary survey report form, and undertaking water-quality analysis. The personnel responsible for data collection in the field therefore need to be trained in a number of skills, including interviewing, working with communities, observation, sampling, and water-quality analysis. Adequate training in these areas will help ensure that surveillance findings are standardized throughout the programme and not subject to regional or local variations.

The importance of training cannot be overemphasized. The training strategies adopted will depend on:

- — the previous training and experience of the staff allocated to surveillance;
- — the range of activities to be undertaken by the surveillance agency and its staff (e.g. hygiene education may or may not be the responsibility of field staff);
- — local water supply practice;
- — the practical organization of surveillance (e.g. whether water-quality testing is to be undertaken on site by field staff or in laboratories).

To ensure that the surveillance agency functions effectively, adequate training should be provided for staff at all levels. Separate training courses are required for field staff, laboratory staff, regional and national managers, and so on. Although not strictly training activities, workshops and seminars for the dissemination of surveillance findings are also important for promotional and motivational reasons.

It is advisable for the surveillance agency to develop a comprehensive strategy for human resource development. This should include clear definition of lines of responsibility and accountability, job descriptions, career structures, and mechanisms for enhancing the motivation of staff at all levels. A training strategy suitable for a four-tier surveillance agency is shown in Fig. 2.3.

For field staff responsible for liaison with communities, on-site water-quality testing, sanitary inspection, and data reporting, the minimum training period should be 2 weeks. This assumes that staff have a general background in environmental health; considerably longer may be required if they have not already received some vocational training. A subject list for a 2-week training course suitable for sanitary technicians working in the field is given in Table 2.2.

***Fig. 2.3 Training strategy for surveillance***

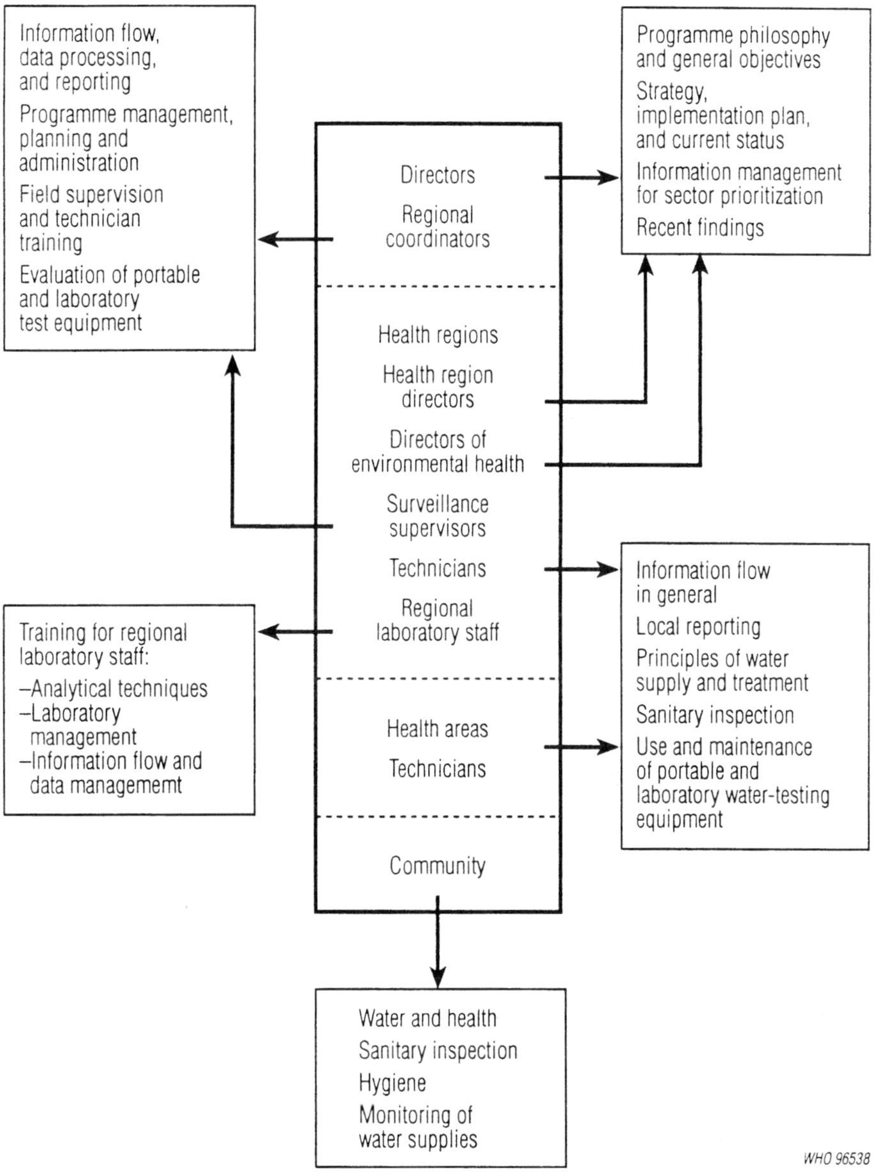

***Table 2.2 Typical subjects for 2-week training module in water-supply surveillance for sanitary technicians***

| |
|---|
| Health, faecal–oral disease transmission |
| Barriers for the control of transmissible diseases |
| Water, food, sanitation, and health education |
| Water-supply surveillance and quality control |
| Water supply: system types and basic characteristics |
| Protection of point-source supplies |
| Gravity-fed systems from protected sources—characteristics and terminology |
| Components of gravity-fed systems from protected sources—points of risk |
| Evaluation and sanitary inspection of gravity-fed systems from protected sources |
| Field visit to gravity-fed system from protected spring source |
| Indicators of faecal contamination, water-quality standards |
| Demonstration of, and laboratory practice with, water-testing equipment |
| Sampling and sample preservation |
| Water-treatment principles |
| Components of rural drinking-water treatment systems |
| Field visit to rural treatment plant, including inspection, sampling, and analysis |
| Information flow in the surveillance programme; reporting |
| Service evaluation (cost, quantity, continuity), water-quality evaluation |
| Fieldwork on evaluation of service quality and water quality in the distribution network |
| Sanitary inspections and inspection report forms |
| Field visit: sanitary inspection and sampling |
| Working with communities |
| Participatory learning techniques |
| Workshop session for planning of implementation activities |
| Round-table discussion |
| Course evaluation |
| Assessment of participants (pre- and post-training) |

Training should not be viewed as a once-only activity, but as a continuous commitment, with follow-up courses, review workshops, and field supervision all contributing to in-service training.

### 2.3.4 Preliminary surveys

The drawing up of work schedules will be determined largely by local conditions such as distance and accessibility (travel time), travel problems of a seasonal nature, and availability of staff, costs, and transport. Targets for minimum frequencies for sanitary inspection and water-quality analysis are given in Chapters 3 and 4 respectively. In many countries even these targets may be difficult to meet in the short term, and they should then be viewed as medium-term goals.

### 2.3.5 Undertaking fieldwork

Information on community surveys and sanitary inspection, and on water sampling and analysis is given in Chapters 3 and 4 respectively. Nevertheless, it is worth considering here two important aspects of field methodology in the context of planning for water-quality surveillance and quality control. Firstly, staff responsible for field activities should ideally give local authorities advance notice of their visit, especially when a representative of the authority concerned must be present to provide access to parts of the supply system; staff should be accompanied by a representative of the supply agency whenever possible. Secondly, after on-site inspection and an analysis of the findings, problems or defects may be pointed out in the field to the local authorities or the representatives of the supply agency.

### 2.3.6 Establishing routine surveillance

When the preliminary survey has been successfully completed, it is possible to plan for routine surveillance. The findings of the preliminary survey may have profound implications for subsequent surveillance activities; for example, surveillance should take due account of the most widely used method of supplying water for domestic purposes or the one that presents the greatest public health risk to the population.

The methods and strategies employed in the preliminary survey should be evaluated and then revised as necessary. This revision should be reflected in training, planning for routine surveillance, expansion of surveillance coverage, surveillance management, and strategies for the promotion of remedial action.

### 2.3.7 Evaluation

Evaluation is an essential stage in the implementation of surveillance programmes. It is greatly assisted by clearly defined project objectives and targets against which progress can be measured. For evaluation to be worth while, it must have clearly defined goals, which should include comparisons with the objectives and targets adopted at the outset. Evaluation should involve personnel from all levels and should result in change when this is indicated. When initial targets have been met, new targets can be defined. A dual-cycle procedure for the evaluation of water-supply surveillance and for promoting and monitoring improvements is illustrated in Fig. 2.4.

## 2.4 Information management

### 2.4.1 Flow and use of information

A general scheme for the flow of information between and within the water-supply and surveillance agencies is shown in Fig. 2.5. Clearly there is an obligation on the part of both agencies to communicate effectively—both laterally and

***Fig. 2.4 Dual-cycle procedure for evaluation of water-supply surveillance***

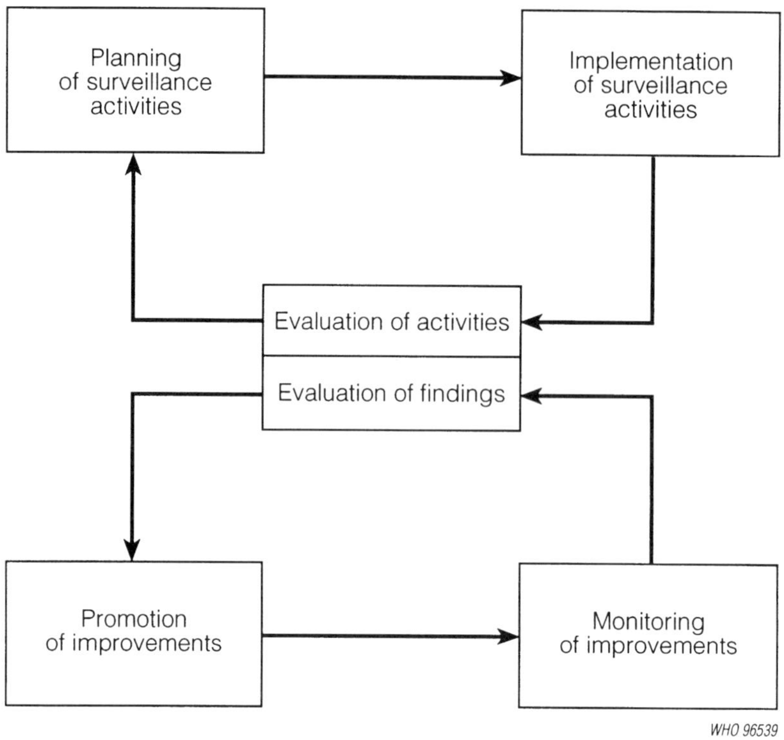


vertically—in order to maximize the quality of service to consumers and to protect public health.

## 2.4.2 Information exchange with suppliers

As the first stage, the surveillance agency should report to the local office of the water supplier. Such reporting should be followed up and, if recommended and feasible corrective action is not undertaken in a reasonable time, notification to a higher authority may be required. The level to which such notifications should be sent may vary according to the supply agency but, in general, it should be the level with ultimate responsibility for the supply.

As a minimum, the information provided by the surveillance agency to the supplier should include:

— a summary of the quality of the service being provided and the condition of the supply; and

— an indication of those aspects considered inadequate and requiring action (with reference to national legislation).

***Fig. 2.5 General scheme for the flow of information between and within the water-supply and surveillance agencies***

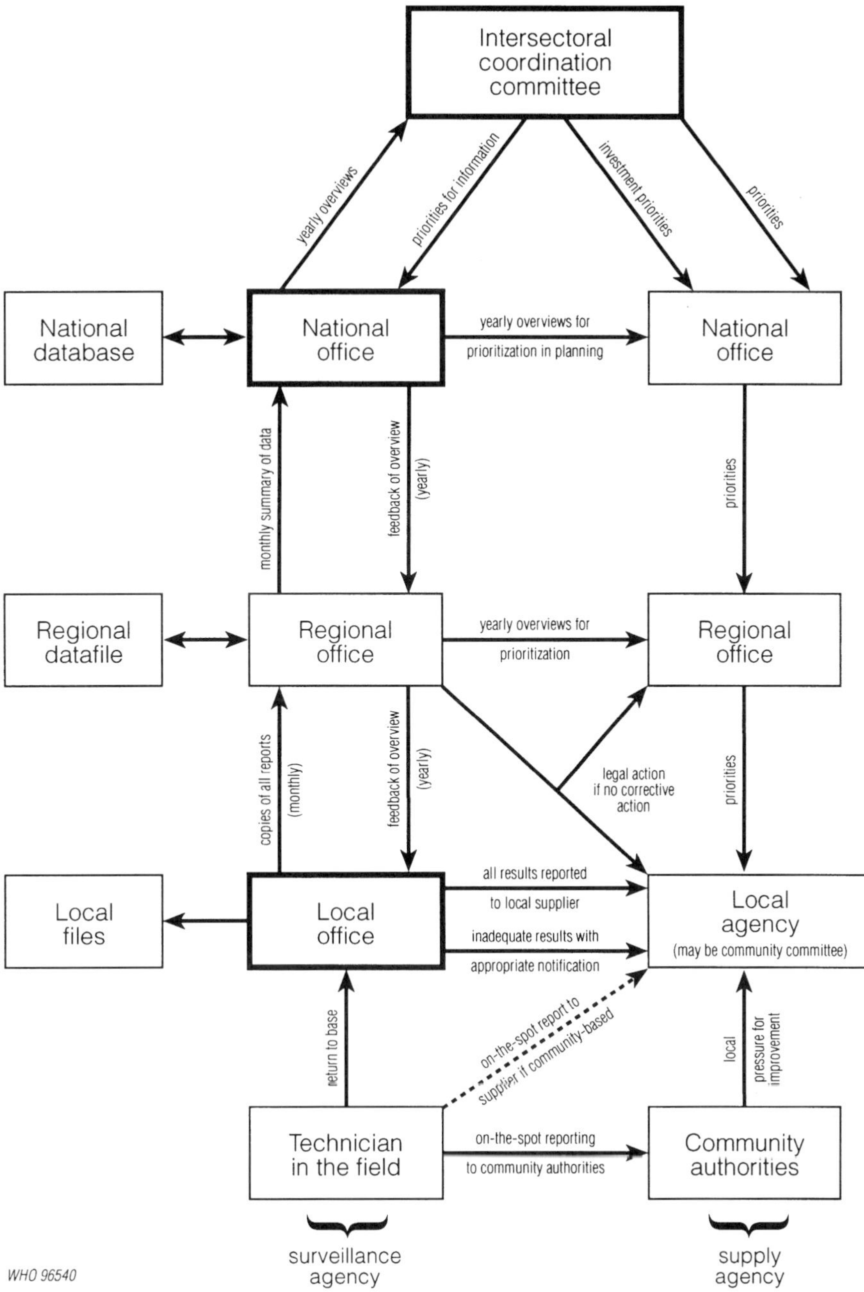

In some circumstances, it may also be necessary to recommend remedial action, such as emergency disinfection of the supply. Some recommendations may require action not only by the supplier, but also by the surveillance agency. For example, if there is severe faecal contamination of drinking-water, this cannot be remedied in the short term, and it may be considered advisable for the population to boil the water. Warning people of the need for this may then be the responsibility of the health authorities (i.e. the surveillance agency). As a further example, if the water supply is of good quality but not continuous, people are forced to store the water in the home, where it may become contaminated; unless a continuous supply can be established quickly, an educational programme on household water storage may be recommended. This is again likely to be the responsibility of the surveillance agency or another agency within the health sector.

The exchange of information between the surveillance and supply agencies should not be limited to complaints about failures. The two agencies must coordinate their activities to ensure regional prioritization, and this requires effective communication and reporting strategy. Reports intended to assist in the setting of regional priorities need not be frequent; annual reporting is likely to be adequate, especially if the report is timed to coincide with the programming of supply agency investments for the following year. Such reports should classify communities and systems in order of priority for intervention based on social and public health criteria. Banding or scores assigned to each community may be used for this purpose (see Chapter 5). Prioritization should not be based solely on water quality, but should also take account of all parameters of drinking-water supply service.

### 2.4.3 Information exchange within the surveillance agency

It is essential that the field worker or local laboratory maintains detailed files on all water supplies in the area. Files should include the results of all inspections and analyses in chronological order. They should be used in conjunction with the inventory, which should include an outline plan of each system, together with details of system components and the population served.

At local level, information is most commonly stored on paper, with perhaps one file per water supply. At regional and national levels, the need for greater data analysis will increasingly justify computerization, although this level of sophistication remains inappropriate at local level in many countries.

The local water-surveillance office should report to each community authority and the relevant supply agency as soon as possible after a field visit. The information should also be passed on to regional authorities to allow follow-up if recommendations for remedial action are not implemented; this may be less urgent and can be done at less frequent intervals, e.g. weekly or monthly. However, there must be a rapid means of reporting in case of emergency.

Regional centres may report to the national surveillance authority quarterly or half-yearly.

### 2.4.4 Information exchange with consumers

The right of consumers to information on health-related parameters of the water supplied to them for domestic purposes is fundamental. However, in many communities, the simple right of access to information will not ensure that individuals are aware of the quality of the water supplied to them; furthermore, the probability of consuming unsafe water is relatively high. The agencies responsible for monitoring should therefore develop strategies for disseminating the health-related results they obtain.

What information is reported to consumers will largely be decided on the basis of the data produced. Nevertheless, raw data (such as the concentrations of contaminants) should be accompanied by some type of interpretation whenever possible, such as compliance or noncompliance with national standards, for the benefit of nonspecialists. The dissemination of information on drinking-water quality must be linked to recommendations for action (e.g. boiling) where appropriate, to community participation in monitoring, and to public education on water-quality issues.

Where reporting incorporates recommendations for remedial action at local level, it may be appropriate to employ pictorial report forms. In some programmes these have been printed alongside the field report forms. In the field, the points that require attention are highlighted, e.g. by circling. The pictorial summary is then torn off and given to the responsible person, together with a full explanation of the actions recommended. Examples of pictorial forms are given in Annex 2.

The delivery of notifications may be difficult, especially in remote communities in developing countries, and methods of solving this problem must therefore be found. Where notifications must be delivered by the monitoring agency itself and distances are considerable, this may become expensive. It may then be more cost-effective to use on-site testing equipment and for field staff to remain in communities overnight. When such staff are adequately trained in the interpretation of results and notification of findings, they can provide an immediate report to the community before returning to their base or proceeding to the next community. However, delayed reporting following sample analysis in a local or regional laboratory remains the most common practice. Suitable means for forwarding reports must then be used. Different means of communication will be found to be appropriate, depending on those available, the urgency of taking corrective action, and the feasibility of implementing the recommended remedial action. Written notification, which may take several days to arrive, will often be adequate, but if urgent action is warranted, e.g. notifying the population of the need to boil its water, more rapid means of communication may be appropriate (telephone, telegram, radio, etc.). Rapid notification should always be followed up by written confirmation.

### 2.4.5 Communication with local and national government

At national level, priorities should be set and disseminated, with recommendations, by means of an annual report. The report should be circulated to all surveillance and supply agencies, the national planning authorities, and agencies involved in coordination within the water-supply sector, e.g. ministries responsible for local government, natural resources, health, and finance. Depending on local circumstances, it may be sent to external support agencies as well, and some nongovernmental organizations may also be recipients. Information exchange with national planning authorities may provide a means of establishing a mutually supportive relationship between surveillance and supply agencies.

Local government should ensure that the agency that supplies drinking-water to the area complies with the surveillance legislation and regulations. Annual reports should be made available which should include information on all breaches of standards and any exemptions or permitted deviations of water quality from national standards. Local government should also actively promote surveillance within the area that it administers, and encourage both producers and consumers to regard surveillance as a positive means of protecting the quality of the water supply.

### 2.4.6 Communication linkage between surveillance and remedial action

Once routine water-supply surveillance activities are established, the links between remedial measures and surveillance should be institutionalized. The most important activities, which should be carried out in the following sequence, are:

- The regional agency responsible for water-supply surveillance prepares an annual plan and fixes a target number of water supplies to be inspected, sampled, and analysed on the basis of inventories.
- Action is coordinated with the community. Sanitary technicians carry out sanitary inspections with community representatives/volunteers. Water samples are analysed on site or transported to a laboratory for analysis.
- The results of the sanitary inspection and water-quality analysis are combined and communicated to the community during the visit if analysis is undertaken on site, or forwarded as soon as possible if samples are processed in a laboratory. In the latter case, the results of the sanitary inspection can be communicated during the visit. The report(s) should indicate the risks identified and the points requiring attention.
- A monthly consolidated report is prepared, covering all points of risk for each facility and the results of analyses.

- The monthly report is transmitted to the regional coordinator, who ranks the relative urgency of action for each facility (see also Chapter 5) and identifies high priorities for remedial action and for hygiene education.
- An urgent action list is sent by the regional coordinator to the appropriate authority for remedial action and to the sanitary technician responsible for monitoring such action.
- Remedial action is taken by the appropriate authority.
- The sanitary technician monitors the remedial action with the community. On completion, he or she repeats the inspection and analysis with the community and communicates the results to the regional coordinator, together with a summary of the remedial work undertaken.
- The coordinator compiles an annual summary of the remedial work undertaken and improvements achieved for review with the supply authorities and by senior staff of the surveillance agency. The report highlights the most common shortcomings, and is used as a basis for identifying the changes in strategy that the supply agency is required to make.
- An annual summary of priorities for hygiene education is compiled by the regional coordinator. A strategy for activities during the following year is agreed with the authority responsible for hygiene education, and the workplan is communicated to the sanitary technicians responsible for surveillance.
- The sanitary technician monitors the hygiene education activities with the community. On completion, he or she evaluates improvements with the community and communicates the results to the regional coordinator, together with a summary of the educational activities undertaken.
- The common shortcomings identified in the annual report are addressed in the supply/construction agencies' annual plans and resources allocated to training, rehabilitation, etc., as appropriate.

### 2.4.7 Use of computers

Data analysis at national level clearly requires the management of large volumes of data, which is a strong argument for computerization. The national agency receives the greatest quantity of data, all of which must be stored, and must also be able to undertake comprehensive data analysis to assist in the setting of priorities at national level.

Where computers are used for data management at national level, it may also be advantageous to extend computerization to the regional centres if they handle sufficient data to warrant it. This has the additional advantage of decentralizing the requirement for data input and reducing the total number of transcriptions, especially if the data are delivered from regional to national centres in computerized form, thereby reducing the chances of error.

At regional level, computerization provides an efficient means of storing information, and possibly also for comparing results with compliance criteria, such as national standards or interim goals agreed with the supplier. The type of communication to be sent to the water supplier will vary according to the nature of the noncompliance, and a computer may also be used to produce appropriate standard letters.

Computerization should not be seen as a universal solution to all problems. As with any other data-management system, the results obtained will be only as good as the data received, and the need for effective data flow and efficient data input is paramount.

## 2.5 Support structure

Ideally, a special section should be established within the responsible agency to oversee and implement activities related to the surveillance programme. This requires a laboratory network, offices, transport, financial support, and adequate staffing.

### 2.5.1 Laboratory network

The laboratory network will vary widely according to a number of criteria. For water-supply surveillance laboratories, the parameters to be measured should be those known to be related to health together with those that may cause water to be rejected by consumers (see section 1.3.1). A laboratory infrastructure may already exist and may include hospital laboratories in the case of surveillance and laboratories at suppliers' water-treatment plants.

In principle, all analyses should be undertaken in a laboratory as close as possible to the site of sampling, taking into account constraints such as staffing and equipment, both of which are largely related to the number of samples analysed and the required frequency of analysis. Prompt analysis minimizes deterioration in sample quality during transport (this is especially important for microbiological samples) and close proximity of the laboratory reduces the costs associated with sample transport.

The range of analyses conducted, the number of samples, and the frequency of sampling may need to be increased progressively with time. The strategy may initially require analysis only of thermotolerant (faecal) coliforms, chlorine residual, and turbidity, before it is expanded to incorporate regional laboratories with a limited analytical range. For quality-control purposes, the range and frequency of analyses may be specified in national standards, but should be increased if conditions deteriorate or if there is any reason to suspect that service quality may be endangered.

A structure based on a central laboratory, a number of regional laboratories, and simple district-level laboratories will almost always be necessary. It may be supplemented by providing field staff with portable equipment for on-site measurement of critical parameters, thus ensuring greater decentralization and more effective coverage (see pp. 65–66).

A central laboratory should be established to undertake a full range of physical, chemical, and microbiological tests. Such laboratories are sometimes referred to as "reference laboratories", although they may not actually perform a reference function. The central laboratory should provide training for analytical staff at all levels, including staff using on-site testing equipment. It should also provide full quality assurance of its own analyses and external quality control for subsidiary laboratories. In addition, it should undertake certain more sophisticated analyses that cannot be decentralized because of the high capital cost of the equipment necessary. These may include, for example, analyses for heavy metals by atomic absorption spectroscopy, and for pesticides by gas chromatography.

Regional laboratories should be able to undertake a moderate range of analyses. They should also provide a support service to remote areas, making culture media and consumables available to staff conducting a limited number of tests using on-site or office-based testing equipment.

Examples of the initial and final laboratory service infrastructure for water-quality analysis are shown in Fig. 2.6.

### 2.5.2 Transport

The preferred means of transport will vary widely depending on the terrain, climatological conditions, and local custom; the possibilities include four-wheel-drive vehicles, pack animals, canoes, bicycles, and motorcycles, in addition to walking (which is always used to some extent). Factors to be taken into account in choosing a means of transport include the need to send samples to the laboratory as quickly as possible (see section 4.1.4), the adequacy of the various forms of transport for the conditions prevailing at the time of surveillance, and price, operating and maintenance costs, and expected useful working life of the transport.

It has been common practice to rely on four-wheel-drive vehicles in carrying out surveillance and quality-control activities in many countries. In some areas, motorcycles have proved particularly successful; they are generally capable of carrying both portable testing equipment and teaching materials, are a far cheaper alternative, and can transport field staff rapidly. They are also less likely to be requisitioned for other purposes.

### 2.5.3 Financial support

Substantial support for surveillance is generally provided by centralized institutions, such as regional or national governments, although they do not cover the total cost. Considerable contributions (which will often be mainly in kind) may also be made by the community itself. Costs may also be reduced by a variety of means, and the water-supply agency should operate on a cost-recovery basis.

***Fig. 2.6 Examples of laboratory service infrastructure***

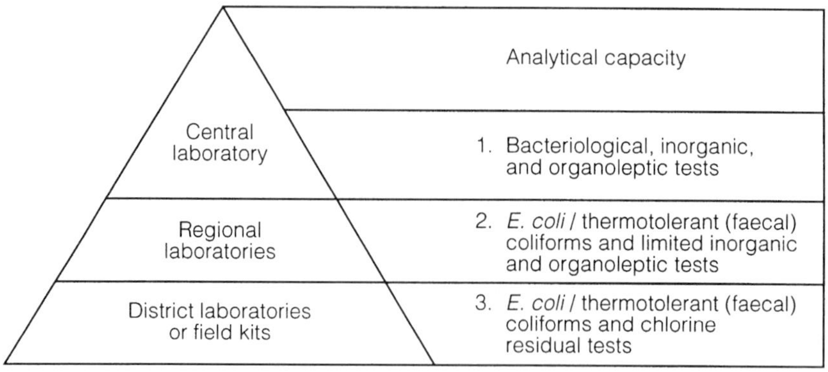


(a) Initial structure

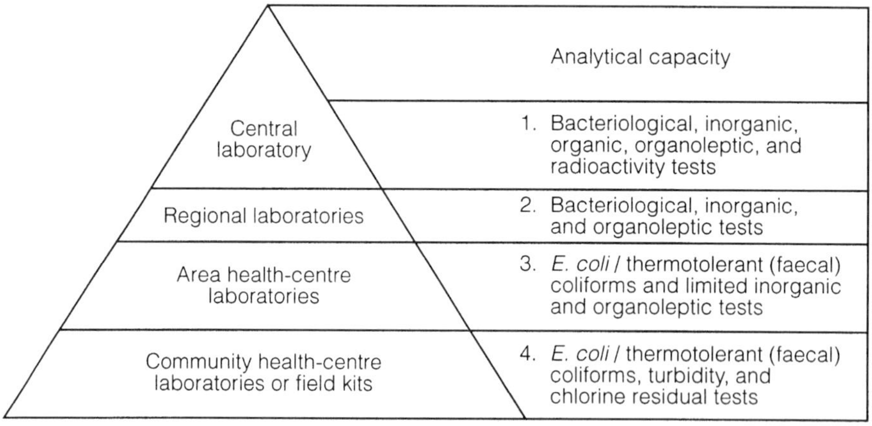


(b) Final structure

WHO 96541

The role and importance of community participation were described in section 1.6. Involving the community in decisions about the establishment of a surveillance programme can be used to create a sense of ownership and a willingness to share some of the costs of surveillance, maintenance, and remedial actions. One approach is to use existing structures in the community, such as water committees, to obtain contributions and to undertake simple maintenance.

A number of strategies may be adopted for minimizing the costs of surveillance. The highest costs are usually those associated with staff and transport, and it is therefore important to concentrate on these. Repeat journeys for resampling or for the delivery of reports to community authorities (essential if there is no suitable postal service or equivalent) are very expensive, and every effort should be made to reduce the need for them. In regions with remote water supplies, this can often be done by using on-site testing equipment, particularly if it is possible for field workers to stay overnight in communities and deliver the results the next morning. If this strategy is adopted, it is often appropriate for field staff to be responsible for health and hygiene education as well as surveillance activities.

### 2.5.4 Staffing

Staffing requirements for servicing a water-supply surveillance programme vary widely, and there is no generally applicable method of determining the number of staff needed for a given population or for a given number of water supplies. The following factors should be borne in mind when staff requirements are estimated:

- Travel to and from water supplies is a major problem for staff undertaking fieldwork (sampling, sanitary inspection, liaison with communities); realistic estimates of travel time should therefore be made at an early stage, and confirmed times entered on the inventory for planning purposes. In addition, seasonal factors such as monsoons may constrain travel at certain times and thus reduce the time available for the work.
- Decentralization of analysis and/or on-site testing becomes more attractive as travel times increase and where water supplies are more widely dispersed.
- The distribution of the workload between point sources, nuclear communities, and piped water supplies will influence the rate at which work is completed.
- The type of supply will also influence the time required, e.g. the sanitary inspection and on-site analysis take an hour in the case of a dug well, while inspection of a piped supply with a source several kilometres away, even for a small community, is likely to take a whole day.
- Greater community involvement will lead to more efficient and effective surveillance, either because it is supported by, and undertaken with, the community or because less frequent visits by the sanitary inspector are necessary.
- Field workers often play an educational role, e.g. in increasing awareness of the health implications of water supply.

Possible responsibilities of surveillance staff at various levels are suggested in Annex 3.

# 3. Surveys

## 3.1 Nature and scope of community surveys

A community survey is an evaluation of all the factors and resources (physical and human) that affect the water-supply service, sanitation, and environmental health of a community. An example of a report form for a community survey is shown in Fig. 3.1; the form will vary with location, and should take account of local conditions.

At the beginning of surveillance programmes, and subsequently at intervals less frequent than those specified (by the surveillance agency) for sanitary inspection, a community survey is required as the foundation of a comprehensive database. The complete community survey should be conducted by the local surveillance agency office (or the area authority in small countries) and should include the following four components:

1. Basic data on water-supply and sanitation facilities with which to update the inventories. Basic inventories have been described in section 2.3.1, and an example of an inventory is shown in Fig. 2.2 on p. 27. The water-supply data (and, in some circumstances, sanitation data) are ideally the responsibility of the water-supply agency; the surveillance agency's field officer should only have to confirm the information. The reality in many countries, however, is that a variety of agencies are involved in water-supply construction, with the result that inventories are often incomplete. The surveillance agency may therefore have to be involved in preparing the inventory.
2. Sanitary inspection (comprising sanitary inspection and water-quality analysis). Sanitary inspection may be conducted by both the water-supply agency and the surveillance agency; the information they generate is shared.
3. A quantitative diagnostic summary of the five key water-supply service indicators (quality, quantity, coverage, continuity, and cost).
4. Hygiene survey. Hygiene surveys are, ideally, the surveillance agency's responsibility.

The quantitative diagnostic summary of water-supply service indicators should be reported to the regional and/or national agency for strategic planning purposes. Figure 3.1 shows a suitable report form. The indicators should be entered into a national database and used to allocate resources for water-supply development and improvement on the basis of priority needs.

## *Fig. 3.1 Example of a community survey form*

District ............................................................ Village ..................................................

Date of visit ........................................................ (Year) Code No. ..................................

Source type ........................................................ System type: ☐ open/closed
☐ piped/unpiped

**Population**

Number of households .........................................

Number of standpipes .........................................

Number of house connections (taps) ...................

School? ...................... School tap? .....................

**Coverage**

$$= \frac{\text{no. standpipes and taps} \times 100\%}{\text{no. households}}$$

= ..............................%

**Flow data**

Flow entering a source ............litres per second

Overflow(s) ..............................litres per second

Flow delivered to taps .............litres per second

**Quantity supplied**

$$= \frac{\text{litres per second delivered} \times 86\,400}{\text{population}}$$

= ............ x 86 400

= .............litres per person per day

**System functioning**

Water entering system on day of inspection?

☐ Yes ☐ No

Proportion of taps operating (tick as appropriate)

| Tap 1 | Tap 2 | Tap 3 |
|---|---|---|
| ☐ Yes ☐ No | ☐ Yes ☐ No | ☐ Yes ☐ No |

Supply all day? ☐ Yes ☐ No

Supply all year? ☐ Yes ☐ No

**Continuity**

$$\text{Overall} = \frac{\text{total 'yes'} \times 100\%}{6} =$$

**Cost**

How many households pay a tariff? .....................

Percentage of household paying .......................%

Monthly tariff ........................................

**Quality**

Thermotolerant (faecal) coliform count per 100 ml

Source count .......................................... /100 ml

Highest count in distribution .................... /100 ml

Sanitary inspection (see accompanying sheet)

Bacteriological quality

Faecal coliform grade: A.B.C.D

Sanitary hazard score

**Remedial action priority** ☐ Low ☐ Intermediate ☐ Urgent

Inspector ............................................................ Signature .............................................

WHO 97005

In Fig. 3.1, the first indicator calculated is *coverage*, which is logical since there is little point in embarking on sanitary inspections until basic infrastructure has been established in a significant proportion of communities. It is also necessary to demonstrate that substantial water-supply coverage has been achieved both in individual communities and in a majority of communities in the district under consideration.

## 3.2 Sanitary inspections

A sanitary inspection is an on-site inspection and evaluation by qualified individuals of all conditions, devices, and practices in the water-supply system that pose an actual or potential danger to the health and well-being of the consumer. It is a fact-finding activity that should identify system deficiencies—not only sources of actual contamination but also inadequacies and lack of integrity in the system that could lead to contamination.

In small communities, where official visits by the surveillance officer are infrequent, it is essential that responsible community members both assist the official in making the survey and learn how to conduct the survey independently. They should sign a report and agree to act on the recommendations where this is feasible (see Chapter 6).

The two principal activities are sanitary inspection and water-quality analysis. It has been suggested that sanitary inspection should take priority over analysis, but the two should be done together wherever possible. They are complementary activities; inspection identifies potential hazards, while analysis indicates whether contamination is occurring and, if so, its intensity.

A sanitary inspection is indispensable for the adequate interpretation of laboratory results. No analytical, bacteriological, or chemical survey, however carefully carried out, is a substitute for comprehensive knowledge of conditions at the water source and within the distribution system, the adequacy of water treatment, and the qualifications and performance of the operators. Samples represent conditions at a single point in time and—even when there is frequent sampling and analysis—the results are reported *after* contamination has occurred, especially in systems without long-term storage. Microbiological contamination is often sporadic and may not be revealed by occasional sampling.

## 3.3 Sanitary inspection reports

The sanitary inspection report is that part of the survey based on the on-site inspection of the water sources (and piped supply systems where appropriate), i.e. a field survey; it therefore provides a direct method of identifying all the *hazards* that are potential and actual causes of contamination of the supply. It is concerned with the physical structure of the supply, its operation, and external environmental factors. The hazards recorded during inspection are often tangible and observable, and may be used together with analytical data to derive a risk assessment.

Sanitary inspections thus provide essential information about immediate and ongoing possible hazards associated with a community water supply, even in the absence of microbiological or chemical evidence of contamination. In addition, inspection of supplies over a period of years provides a longer-term perspective and assists in the identification and minimization of risks caused by progressive deterioration in any aspect of the supply.

### 3.3.1 Functions of sanitary inspection report forms

Inspection forms should provide a simple and rapid means of assessing and identifying hazards associated with water-supply systems. Wherever sanitary inspections are carried out, there will inevitably be a variety of systems to consider, and a decision must then be made on whether to attempt to produce a single inspection form that deals with all types of system or to produce a series of forms, each dealing with a different type. Some of the information that it may be useful to include on one inspection form may already have been collected for inventory purposes. Again, a decision must be made on how much of this kind of detail it is appropriate to include.

The inspection form should include at least a checklist of the components of the water supply from source to distribution and incorporate all the potential points where hazards may be introduced. Any problems identified during the inspection should be highlighted so that a report may be provided directly to the community and copies forwarded to both supply agency and health authority.

The specific functions of the sanitary inspection report are to:

- identify potential sources and points of contamination of the water supply;
- quantify the hazard (*hazard score*) attributable to the sources and supply;
- provide a clear, graphical means of explaining the hazards to the operator/user;
- provide clear guidance as to the remedial action required to protect and improve the supply;
- provide the raw data for use in systematic, strategic planning for improvement.

The sanitary inspection report may be considered as an integral part of a community survey as defined in section 3.1. It should therefore not be restricted to factors that may cause problems with water quality, but should also take into account other service indicators, e.g. coverage, cost, continuity, and quantity. This is particularly important for supply agencies that may wish to give special consideration to such factors from the point of view of operation and maintenance. It should be possible to determine an overall measure of the sanitary state of the supply based on the checklist, and this hazard or risk score may be used in deciding priorities for remedial action by the community or by whichever agency is best able to intervene and make improvements.

### 3.3.2 Design of sanitary inspection report forms

The design, evaluation, and refinement of sanitary inspection forms are among the most important aspects of developing a surveillance or quality-control programme. Two approaches are possible—the use of pictures and brief checklists, or the use of more detailed checklists with explanatory notes. Either may be used successfully. However, in some countries where the level of training of environmental health inspectors or sanitary technicians may not be very high, the use of pictorial inspection forms may be the most effective method, and is therefore considered here.

Ideally, forms should be designed in such a way that the community or owner of the supply can either conduct the survey or be given a summary of the problems identified before the departure of the inspector. This means that any actions required at local level can be agreed and implementation can be started immediately. Where actions are required by others, e.g. water-supply or health agencies, the community should also be informed of the recommendations that will be made. Copies of the full sanitary survey should be sent to all relevant authorities, and this is facilitated by well designed inspection forms, for example with duplicate or triplicate sheets and "tear-off" slips for recommended actions.

A series of model inspection report forms is presented in Annex 2. With one exception, these are in double-page format and include illustrations of a range of water supplies in a recognizable setting; potential hazards are identified and numbered. The forms include details of the type of facility, the supply, the date of the sanitary inspection visit, and so on. The checklist of 10 or more points allows a hazard score to be assigned based on the total number of hazards identified.

In some countries it may be necessary to consider hazards other than those illustrated in Annex 2, and these should also be included in the checklist. Sanitary inspection forms should be designed to match local circumstances; they should be suitable for the inspectors to use, and the recipients of the information should be able to understand and act on them. Any pictures that are included must be carefully drawn to reflect the cultures and situations that they are designed to depict. The range of report forms given in Annex 2 covers most of the main types of small water-supply installations. Nevertheless, the list is not exhaustive, and local variations in design and in cultural habits may have a profound impact on the design of such forms.

The principle on which the design of sanitary inspection report forms is based is that every fault that may reduce the quality of the supply should be listed and checked during the sanitary inspection. Each fault represents a sanitary hazard. Every additional fault increases the probability that contamination will occur; the number of hazards may therefore be totalled to provide an additive sanitary risk score, but this implies an equal weighting of all the risks. However, it is most unlikely that such equal weighting will be correct and that the score will be directly proportional to the intensity of the resulting contamination. Thus it is important to incorporate differential weighting for local conditions that permits a better interpretation of the information and promotes remedial action.

## 3.4 Carrying out sanitary inspections

Staff responsible for field sanitary inspection work should always try to *notify the local community representatives in advance of the visit*, especially where the presence of the latter is required in order to obtain access to certain points in the supply system and where the assistance of community members in conducting the inspection is needed.

On arrival in the community, the surveillance officer must verify basic data with community representatives, as indicated in Chapter 2 (Fig. 2.2). Any records that the community keeps, for example of tariffs, should be examined and the information noted, including the amount charged and the number of households paying.

Before visiting the community, the surveillance officer may have prior knowledge of the type and number of supplies, sources, and taps. This should be checked against local records and maps held by the local health post or health centre, for example. If no map is available, an attempt should be made to prepare at least a sketch map of the supply or sources.

Much of the information required for the investigation of drinking-water supply services will be obtained by interviewing community members; this is especially important when visiting households to assess the continuity of service. Wherever possible, the surveillance officer should verify any information so obtained by direct observation during the field survey.

While it may appear logical for inspection and sampling to begin at the source of piped supply and to progress through the system with the flow of water, the converse is actually the case. Working against the flow (i.e. beginning with the distribution network and progressing up through the system) makes it less likely that any samples taken will have been accidentally contaminated by the sampler earlier in the system, e.g. when opening little-used lids of reservoirs or protected spring sources.

The surveillance officer should complete the sanitary inspection report on site together with the community representatives. Opportunities to point out problems or defects in the field to community members, their representatives, or the system caretaker or operator should be taken whenever possible. It may also be appropriate to undertake simple repairs, e.g. replacement of washers in public taps, at the same time.

After completing the sanitary inspection, the survey officer should circle each of the points of risk on the diagram, preferably in red ink. The diagram (see Annex 2) should be separated from the inspection report form and given to a member of the water committee or community representative. Before leaving the community, the surveillance officer should discuss, agree, and schedule any follow-up actions and indicate the date of the next survey.

The survey officer carrying out the sanitary survey should record whether or not sampling or analysis will be undertaken. Labour and hence time can sometimes be saved by carrying out the analysis in the field at the same time as the inspection; elsewhere, water analysis may be part of follow-up, with samples transported to a laboratory for testing.

Some countries have introduced special postcards the community can use to report serious operational or remedial requirements; these are posted to the agency responsible for operation and maintenance, which then makes an appropriate response and provides the necessary technical support.

The procedural steps for carrying out a sanitary survey are summarized in Fig. 3.2.

## 3.5 Timing and frequency of sanitary inspections

Sanitary inspections should be undertaken on a regular basis, ideally at the frequencies indicated in Table 3.1.

### 3.5.1 New sources

One of the most important surveys is that undertaken when new sources of water are being developed. This survey should provide sufficient information to indicate the suitability of the source and the amount of treatment required before the water can be considered suitable for human consumption. When alternative water sources are under consideration, each should be surveyed. Physical, bacteriological, and chemical analyses should be carried out during catchment surveys when new water sources are explored to assess potential new water supplies. Chemical and bacteriological analyses should also be done when hydrogeographical surveys are carried out. The guiding principle is that *no new public water supply should be approved without a sanitary inspection.*

Surface-water sources may be extremely difficult to survey adequately, particularly in remote rural areas and where land-use patterns are changing rapidly. Not only may there be daily and seasonal changes in flow to consider but, in addition, variations in physical, chemical, and microbiological characteristics necessitate analysis throughout the year to take account of the effect of changes in rainfall patterns.

### 3.5.2 Routine surveys of existing supplies

Although it is unrealistic in most instances to expect the surveillance agency to devote more than 1 or 2 days per system each year to a survey, this can hardly be

**Fig. 3.2 Procedural steps for carrying out a sanitary survey**

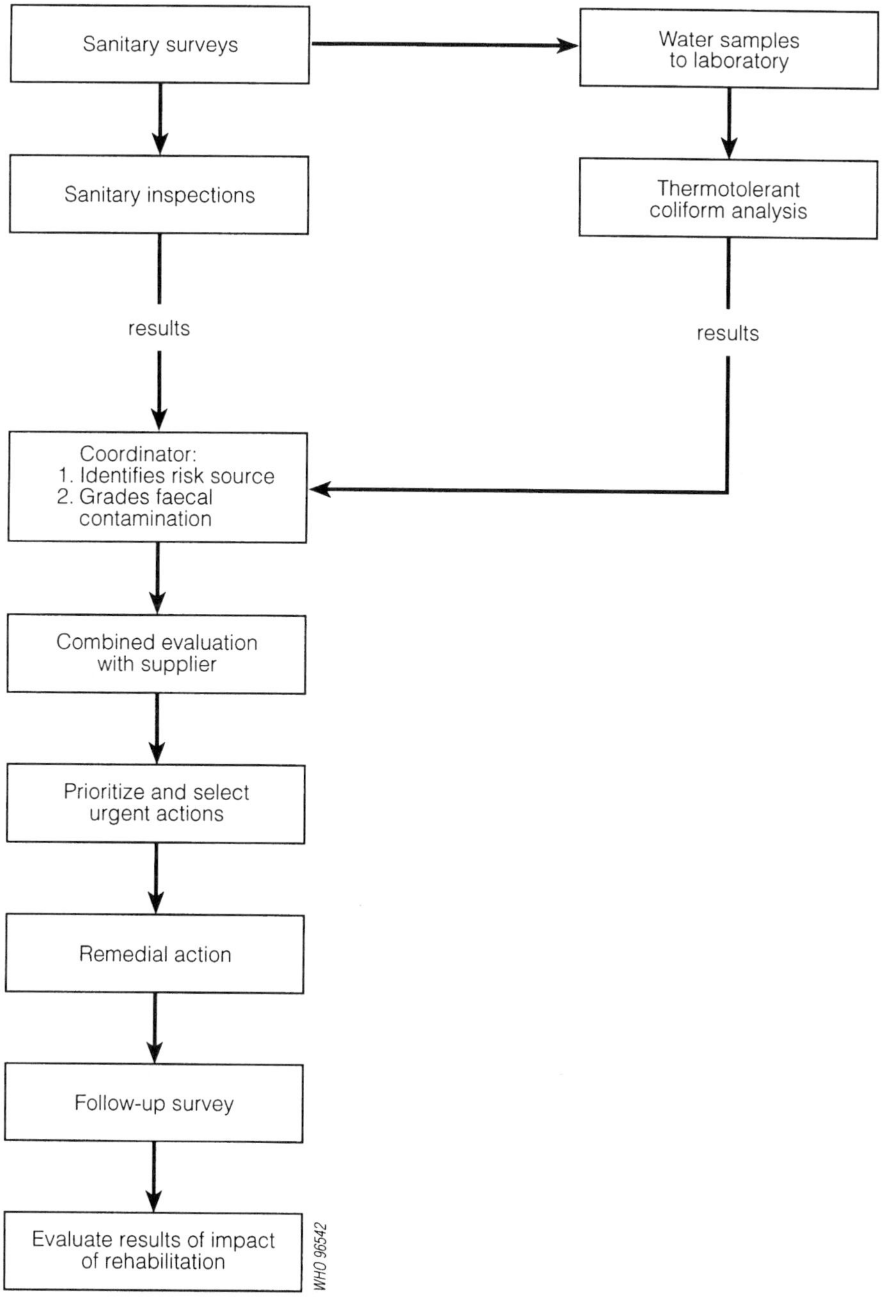

***Table 3.1 Suggested minimum annual frequency of sanitary inspections***

| Source and mode of supply | Community[a] | Water-supply agency[b] | Surveillance agency[a,b,c] |
|---|---|---|---|
| Dug well (without windlass) | 6 | — | 1[d] |
| Dug well (with windlass) | 6 | — | 1[d] |
| Dug well with hand-pump | 4 | — | 1[d] |
| Shallow and deep tubewell with hand-pump | 4 | — | 1[d] |
| Rainwater catchment | 4 | — | 1[d] |
| Gravity spring | 4 | — | 1[d] |
| Piped supply: groundwater sources (springs and wells), with and without chlorination | — | 1 | 1 |
| Treated surface source of piped supply, with chlorination: | | | |
| <5000 population | 12 | 1 | 1 |
| 5000–20 000 population | — | 2 | 1 |
| Distribution system of piped supply[e] | — | 12 | 1 |

[a] For family-owned facilities (e.g. dug wells with or without hand-pumps), the family is responsible for inspections, with support from the surveillance agency.
[b] All new sources should be inspected before commissioning.
[c] Under emergency conditions, such as onset of epidemic diseases, inspection should take place immediately.
[d] Where it is impractical to inspect all such facilities, a statistically significant sample should be inspected.
[e] Public standposts are cleaned by the community if the population is less than 5000. The water-supply agency maintains the distribution system and tapstands if the population is between 5000 and 20 000.

considered adequate. Thus, sanitary surveys should also be undertaken periodically by water-supply agency staff as well as by the surveillance agency.

Traditionally, the frequency of inspection and analyses has been based on population size. For community supplies, it is necessary to involve community members, especially where there is no official water-supply agency. The diversity of water-supply facilities and administrative arrangements makes it difficult to provide other than general guidelines for the frequency of these surveys, as suggested in Table 3.1. However, it is important to note that these suggested frequencies are minimum values. It is also vital that any community report which suggests that serious risks exist should be officially logged and acknowledged, and that follow-up action is taken by the surveillance agency.

# 4. Water sampling and analysis

Ideally, a laboratory infrastructure should be established which will enable all samples to be returned to a central or regional laboratory within a few hours of being taken. However, this depends on the availability of a good road system and of reliable motorized transport for all sampling officers, and these are not available in many countries. Thus, although it may be possible to establish well-equipped central and even regional laboratories for water analysis, at the provincial and district levels it may be necessary to rely on a relatively small number of simple tests. As noted in Chapter 1, this approach is sometimes called critical-parameter water testing.

The most important factor to take into account is that, in most communities, the principal risk to human health derives from faecal contamination. In some countries there may also be hazards associated with specific chemical contaminants such as fluoride or arsenic, but the levels of these substances are unlikely to change significantly with time. Thus, if a full range of chemical analyses is undertaken on new water sources and repeated thereafter at fairly long intervals, chemical contaminants are unlikely to present an unrecognized hazard. In contrast, the potential for faecal contamination in untreated or inadequately treated community supplies is always present. The minimum level of analysis should therefore include testing for indicators of faecal pollution (thermotolerant (faecal) coliforms), turbidity, and chlorine residual and pH (if the water is disinfected with chlorine).

Even in developing countries poorly served by roads and transportation, it is usually possible to devise a rational sampling and analytical strategy. This should incorporate carefully selected critical-parameter tests in remote (usually rural) locations using simple methods and portable water-testing equipment (see pp. 65–66) where appropriate. Wherever possible the community should be involved in the sampling process. Where water is disinfected, primary health workers, schoolteachers, and sometimes community members can be trained to carry out simple chlorine residual testing. The same people could also collect samples for physicochemical analysis and arrange for their delivery to the regional laboratory. The use of community members in this way has significant implications for training and supervision but would be one way of ensuring more complete surveillance coverage.

## 4.1 Sampling

The guidelines provided here take into account experience in surveillance programmes in remote, typically rural, areas and in periurban communities. More general advice on sampling is given in Volume 1 and in ISO standards (see the Bibliography).

### 4.1.1 Location of sampling points

One objective of surveillance is to assess the quality of the water supplied by the supply agency and of that at the point of use, so that samples of both should be taken. Any significant difference between the two has important implications for remedial strategies.

Samples must be taken from locations that are representative of the water source, treatment plant, storage facilities, distribution network, points at which water is delivered to the consumer, and points of use. In selecting sampling points, each locality should be considered individually; however, the following general criteria are usually applicable:

- Sampling points should be selected such that the samples taken are representative of the different sources from which water is obtained by the public or enters the system.
- These points should include those that yield samples representative of the conditions at the most unfavourable sources or places in the supply system, particularly points of possible contamination such as unprotected sources, loops, reservoirs, low-pressure zones, ends of the system, etc.
- Sampling points should be uniformly distributed throughout a piped distribution system, taking population distribution into account; the number of sampling points should be proportional to the number of links or branches.
- The points chosen should generally yield samples that are representative of the system as a whole and of its main components.
- Sampling points should be located in such a way that water can be sampled from reserve tanks and reservoirs, etc.
- In systems with more than one water source, the locations of the sampling points should take account of the number of inhabitants served by each source.
- There should be at least one sampling point directly after the clean-water outlet from each treatment plant.

Sampling sites in a piped distribution network may be classified as:

— fixed and agreed with the supply agency;

— fixed, but not agreed with the supply agency; or

— random or variable.

Each type of sampling site has certain advantages and disadvantages. Fixed sites agreed with the supplier are essential when legal action is to be used as a means of ensuring improvement; otherwise, the supply agency may object to a sample result on the grounds that water quality may have deteriorated in the household, beyond the area of responsibility of the supplier. Nevertheless, fixed sample points are rare or unknown in some countries.

Fixed sites that are not necessarily recognized by the supply agency are used frequently in investigations, including surveillance. They are especially useful when results have to be compared over time, but they limit the possibility of identifying local problems such as cross-connections and contamination from leaking distribution networks.

Sampling regimes using variable or random sites have the advantage of being more likely to detect local problems but are less useful for analysing changes over time.

### 4.1.2 Sampling frequency

The most important tests used in water-quality surveillance or quality control in small communities are those for microbiological quality (by the measurement of indicator bacteria) and turbidity, and for free chlorine residual and pH where chlorination is used. These tests should be carried out whenever a sample is taken, regardless of how many other physical or chemical variables are to be measured. The recommended minimum frequencies for these critical measurements in unpiped water supplies are summarized in Table 4.1 and minimum sample numbers for piped drinking-water in the distribution system are shown in Table 4.2.

### 4.1.3 Sampling methods for microbiological analysis

Detailed methods for sampling for microbiological analysis are given in Annex 4. All samples should be accompanied by an appropriate collection form; a model sample collection form is illustrated in Fig. 4.1.

### 4.1.4 Storage of samples for microbiological analysis

Although recommendations vary, the time between sample collection and analysis should, in general, not exceed 6 hours, and 24 hours is considered the absolute maximum. It is assumed that the samples are immediately placed in a lightproof insulated box containing melting ice or ice-packs with water to ensure rapid cooling. If ice is not available, the transportation time must not exceed 2 hours. It is imperative that samples are kept in the dark and that cooling is rapid. If these conditions are not met, the samples should be discarded. When water that contains or may contain even traces of chlorine is sampled, the chlorine must be inactivated. If it is not, microbes may be killed during transit and an erroneous result will be obtained. The bottles in which the samples are placed should therefore contain sodium thiosulfate to neutralize any chlorine present, as de-

**Table 4.1 Minimum frequency of sampling and analysis of unpiped water supplies**

| Source and mode of supply | Minimum frequency of sampling and analysis | | Remarks |
|---|---|---|---|
| | **Bacteriological** | **Physical/chemical** | |
| Open wells for community supply | Sanitary protection measures; bacteriological testing only if situation demands | Once initially for community wells | Pollution usually expected to occur |
| Covered dug wells and shallow tubewells with hand-pumps | Sanitary protection measures; bacteriological testing only if situation demands | Once initially, thereafter as situation demands | Situations requiring testing: change in environmental conditions, outbreak of waterborne disease, or increase in incidence of waterborne diseases |
| Deep tubewells with hand-pumps | Once initially, thereafter as situation demands | Once initially, thereafter as situation demands | Situations requiring testing: change in environmental conditions, outbreak of waterborne disease, or increase in incidence of waterborne diseases |
| Protected springs | Once initially, thereafter as situation demands | Periodically for residual chlorine if water is chlorinated | Situations requiring testing: change in environmental conditions, outbreak of waterborne disease, or increase in incidence of waterborne diseases |
| Community rainwater collection systems | Sanitary protection measures; bacteriological testing only if situation demands | Not needed | — |

## *Fig. 4.1 Model sample collection form*

WATER-QUALITY
CONTROL
PROGRAMME

[Name of body responsible. . . . . . . . . . . . . . . . . . . . . . . . . . . . . ]

SAMPLING AND BACTERIOLOGICAL ANALYSIS

Sampling data:

Locality __________

Sample site __________

Place __________

Source __________

Sender __________

Date collected __________

Time collected __________

Date of analysis __________

Time of analysis __________

Residual chlorine __________ mg/litre

Results:

TOTAL COLIFORMS . . . . . . . . . . . . __________ /100 ml

FAECAL COLIFORMS . . . . . . . . . . . __________ /100 ml

(OTHER) . . . . . . . . . . . . . . . . . . . . . __________

Laboratory Sample No. __________

WATER BACTERIOLOGICALLY

GOOD – BAD

ACTION TAKEN

__________

__________

__________

__________
(signed)

SAMPLE DATA

Locality __________

Sample site __________

Place __________

Source __________

Residual chlorine __________

Date of sampling __________

Time of sampling __________

Sender __________

Section detached and stuck to the sample bottle

Analytical results entered by laboratory; copies of this section sent by laboratory to local surveillance agency or water-supply agency and person responsible for sampling

WHO 96543

***Table 4.2 Minimum sample numbers for piped drinking-water in the distribution system***

| Population served | No. of monthly samples |
|---|---|
| <5000 | 1 |
| 5000–100 000 | 1 per 5000 population |
| >100 000 | 1 per 10 000 population, plus 10 additional samples |

scribed in Annex 4. The box used to carry samples (see Fig. 4.2) should be cleaned and disinfected after each use to avoid contaminating the surfaces of the bottles and the sampler's hands.

### 4.1.5 Sampling methods for physicochemical analysis

Results of physicochemical analysis are of no value if the samples tested are not properly collected and stored. This has important consequences for sampling regimes, sampling procedures, and methods of sample preservation and storage. In general, the time between sampling and analysis should be kept to a minimum. Storage in glass or polyethylene bottles at a low temperature (e.g. 4 °C) in the dark is recommended. Sample bottles must be clean but need not be sterile. Special preservatives may be required for some analytes. Residual chlorine, pH, and turbidity should be tested immediately after sampling as they will change during storage and transport.

## 4.2 Bacteriological analysis

The principal risk associated with water in small-community supplies is that of infectious disease related to faecal contamination. Hence, as described in Chapter 1, the microbiological examination of drinking-water emphasizes assessment of the hygienic quality of the supply. This requires the isolation and enumeration of organisms that indicate the presence of faecal contamination. In certain circumstances, the same indicator organisms may also be used to assess the efficiency of drinking-water treatment plants, which is an important element of quality control. Other microbiological indicators, not necessarily associated with faecal pollution, may also be used for this purpose.

The isolation of specific pathogens in water should be undertaken only by reference laboratories for purposes of investigating and controlling outbreaks of disease. Routine isolation in other circumstances is not practical.

Detailed methods for use in bacteriological analysis are described in Annex 5 (multiple-tube method), Annex 6 (membrane-filtration method), Annex 7 (on-site testing method), and Annex 8 (presence–absence test).

***Fig. 4.2 Transport box for samples for microbiological analysis***

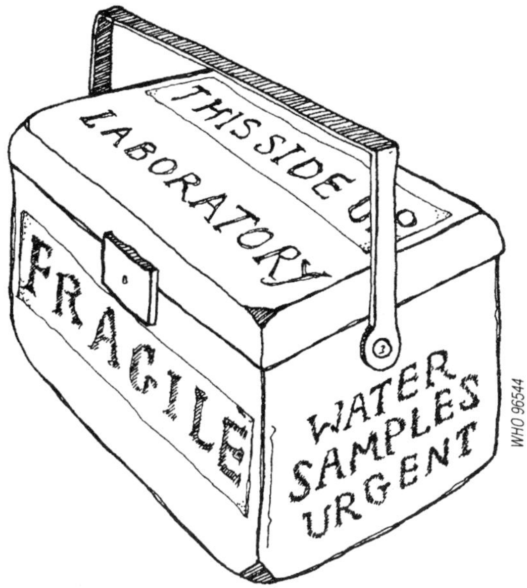

Ice-pack or
freezing mixture

### 4.2.1 Indicator organisms

The properties and significance of the commonly used faecal indicator bacteria are described in detail in Volume 1; a summary is provided here.

*Escherichia coli* is a member of the family Enterobacteriaceae, and is characterized by possession of the enzymes β-galactosidase and β-glucuronidase. It grows at 44–45 °C on complex media, ferments lactose and mannitol with the production of acid and gas, and produces indole from tryptophan. However, some strains can grow at 37 °C but not at 44–45 °C, and some do not produce gas. *E. coli* does not produce oxidase or hydrolyse urea. Complete identification of the organism is too complicated for routine use, but a number of tests have been developed for rapid and reliable identification. Some of these methods have been standardized at international and national levels and accepted for routine use; others are still being developed or evaluated.

*Escherichia coli* is abundant in human and animal faeces; in fresh faeces it may attain concentrations of $10^9$ per gram. It is found in sewage, treated effluents, and all natural waters and soils subject to recent faecal contamination, whether from humans, wild animals, or agricultural activity. Recently, it has been suggested that *E. coli* may be present or even multiply in tropical waters not subject to human faecal pollution. However, even in the remotest regions, faecal contamination by wild animals, including birds, can never be excluded. Because animals can transmit pathogens that are infective in humans, the presence of *E. coli* or thermotolerant coliform bacteria must not be ignored, because the presumption remains that the water has been faecally contaminated and that treatment has been ineffective.

#### *Thermotolerant coliform bacteria*

Thermotolerant coliform bacteria are the coliform organisms that are able to ferment lactose at 44–45 °C; the group includes the genus *Escherichia* and some species of *Klebsiella*, *Enterobacter*, and *Citrobacter*. Thermotolerant coliforms other than *E. coli* may also originate from organically enriched water such as industrial effluents or from decaying plant materials and soils. For this reason, the term "faecal" coliforms, although frequently employed, is not correct, and its use should be discontinued.

Regrowth of thermotolerant coliform organisms in the distribution system is unlikely unless sufficient bacterial nutrients are present, unsuitable materials are in contact with the treated water, the water temperature is above 13 °C, and there is no free residual chlorine.

In most circumstances, concentrations of thermotolerant coliforms are directly related to that of *E. coli*. Their use in assessing water quality is therefore considered acceptable for routine purposes, but the limitations with regard to specificity should always be borne in mind when the data are interpreted. If high counts of thermotolerant coliforms are found in the absence of detectable sanitary hazards, additional confirmatory tests specific for *E. coli* should be carried out. National reference laboratories developing national standard methods are advised to examine the specificity of the thermotolerant coliform test for *E. coli* under local conditions.

Because thermotolerant coliform organisms are readily detected, they have an important secondary role as indicators of the efficiency of water-treatment processes in removing faecal bacteria. They may therefore be used in assessing the degree of treatment necessary for waters of different quality and for defining performance targets for removal of bacteria.

### Coliform organisms (total coliforms)

Coliform organisms have long been recognized as a suitable microbial indicator of drinking-water quality, largely because they are easy to detect and enumerate in water. The term "coliform organisms" refers to Gram-negative, rod-shaped bacteria capable of growth in the presence of bile salts or other surface-active agents with similar growth-inhibiting properties and able to ferment lactose at 35–37 °C with the production of acid, gas, and aldehyde within 24–48 hours. They are also oxidase-negative and non-spore-forming and display β-galactosidase activity.

Traditionally, coliform bacteria were regarded as belonging to the genera *Escherichia*, *Citrobacter*, *Enterobacter*, and *Klebsiella*. However, as defined by modern taxonomical methods, the group is heterogeneous. It includes lactose-fermenting bacteria, such as *Enterobacter cloacae* and *Citrobacter freundii*, which can be found in both faeces and the environment (nutrient-rich waters, soil, decaying plant material) as well as in drinking-water containing relatively high concentrations of nutrients, as well as species that are rarely, if ever, found in faeces and may multiply in relatively good-quality drinking-water, e.g. *Serratia fonticola*, *Rabnella aquatilis*, and *Buttiauxella agrestis*.

The existence both of non-faecal bacteria that fit the definitions of coliform bacteria and of lactose-negative coliform bacteria limits the applicability of this group as an indicator of faecal pollution. Coliform bacteria should not be detectable in treated water supplies and, if found, suggest inadequate treatment, post-treatment contamination, or excessive nutrients. The coliform test can therefore be used as an indicator both of treatment efficiency and of the integrity of the distribution system. Although coliform organisms may not always be directly related to the presence of faecal contamination or pathogens in drinking-water, the coliform test is still useful for monitoring the microbial quality of treated piped water supplies. If there is any doubt, especially when coliform organisms are found in the absence of thermotolerant coliforms and *E. coli*, identification to the species level or analyses for other indicator organisms may be undertaken to investigate the nature of the contamination. Sanitary inspections will also be needed.

### *Faecal streptococci*

Faecal streptococci are those streptococci generally present in the faeces of humans and animals. All possess the Lancefield group D antigen. Taxonomically, they belong to the genera *Enterococcus* and *Streptococcus*. The taxonomy of enterococci has recently undergone important changes, and detailed knowledge of the ecology of many of the new species is lacking; the genus *Enterococcus* now includes all streptococci that share certain biochemical properties and have a wide tolerance of adverse growth conditions—*E. avium*, *E. casseliflavus*, *E. cecorum*, *E. durans*, *E. faecalis*, *E. faecium*, *E. gallinarum*, *E. hirae*, *E. malodoratus*, *E. mundtii*, and *E. solitarius*. Most of these species are of faecal origin and can generally be regarded as specific indicators of human faecal pollution for most practical purposes. They may, however, be isolated from the faeces of animals, and certain species and subspecies, such as *E. casseliflavus*, *E. faecalis* var. *liquefaciens*, *E. malodoratus*, and *E. solitarius*, occur primarily on plant material.

In the genus *Streptococcus*, only *S. bovis* and *S. equinus* possess the group D antigen and therefore belong to the faecal streptococcus group. They derive mainly from animal faeces. Faecal streptococci rarely multiply in polluted water, and they are more persistent than *E. coli* and coliform bacteria. Their primary value in water-quality examination is therefore as additional indicators of treatment efficiency. Moreover, streptococci are highly resistant to drying and may be valuable for routine control after new mains are laid or distribution systems are repaired, or for detecting pollution of groundwaters or surface waters by surface run-off.

## 4.2.2 Principal analytical techniques

The standardization of methods and laboratory procedures is important. International standard methods should be evaluated under local conditions before they are formally adopted by national surveillance programmes. A list of ISO standard methods is given in the Bibliography. The methods described in the annexes to this publication are based on these ISO standard methods, modified where appropriate in the light of experience in the surveillance of community water supplies.

The principal methods used in the isolation of indicator organisms from water are the membrane-filtration (MF) method, the multiple-tube (MT) or most probable number (MPN) method and presence–absence tests.

### *Membrane-filtration method*

In the membrane-filtration (MF) method, a minimum volume of 10 ml of the sample (or dilution of the sample) is introduced aseptically into a sterile or properly disinfected filtration assembly containing a sterile membrane filter (nominal pore size 0.2 or 0.45 μm). A vacuum is applied and the sample is drawn

***Table 4.3 Typical sample volumes for membrane-filtration analysis***

| Sample type | Sample volume (ml) |
|---|---|
| Treated drinking-water | 100 |
| Partially treated drinking-water | 10–100 |
| Protected source water or groundwater | 10–100 |
| Surface water and water from open wells | 0.1–100[a] |

[a] Volumes less than 10 ml should be added to the filtration apparatus after addition of at least 10 ml of sterile diluent to ensure adequate dispersal across the surface of the membrane filter.

through the membrane filter. All indicator organisms are retained on or within the filter, which is then transferred to a suitable selective culture medium in a Petri dish. Following a period of resuscitation, during which the bacteria become acclimatized to the new conditions, the Petri dish is transferred to an incubator at the appropriate selective temperature where it is incubated for a suitable time to allow the replication of the indicator organisms. Visually identifiable colonies are formed and counted, and the results are expressed in numbers of "colony-forming units" (CFU) per 100 ml of original sample.

This technique is inappropriate for waters with a level of turbidity that would cause the filter to become blocked before an adequate volume of water had passed through. When it is necessary to process low sample volumes (less than 10 ml), an adequate volume of sterile diluent must be used to disperse the sample before filtration and ensure that it passes evenly across the entire surface of the membrane filter. Membrane filters may be expensive in some countries.

Typical sample volumes for different water types are shown in Table 4.3. Where the quality of the water is totally unknown, it may be advisable to test two or more volumes in order to ensure that the number of colonies on the membrane is in the optimal range for counting (20–80 colonies per membrane).

### *Multiple-tube method*

The multiple-tube method is also referred to as the most probable number (MPN) method because—unlike the MF method—it is based on an indirect assessment of microbial density in the water sample by reference to statistical tables to determine the most probable number of microorganisms present in the original sample. It is essential for highly turbid samples that cannot be analysed by membrane filtration. The technique is used extensively for drinking-water analysis, but it is time-consuming to perform and requires more equipment, glassware, and consumables than membrane filtration. However, the multiple-tube method may be more sensitive than membrane filtration.

**Table 4.4 *Typical sample volumes and numbers of tubes for the multiple-tube method***

| Sample type | Number of tubes for sample volume: | | | | |
|---|---|---|---|---|---|
| | 50 ml | 10 ml | 1 ml | 0.1 ml | 0.01 ml[a] |
| Treated drinking-water | 1 | 5 | — | — | — |
| Partially treated drinking-water | — | 5 | 5 | 5 | — |
| Protected source water or groundwater | — | 5 | 5 | 5 | — |
| Surface water or water from open wells | — | — | 5 | 5 | 5 |

[a] Volumes of 0.1 and 0.01 ml are tested by the addition of 1 ml of a 1/10 and 1/100 dilution sample, respectively, to 10 ml of single-strength culture medium.

The multiple-tube method depends on the separate analysis of a number of volumes of the same sample. Each volume is mixed with culture medium and incubated. The concentration of microorganisms in the original sample can then be estimated from the pattern of positive results (the number of tubes showing growth in each volume series) by means of statistical tables that give the "most probable number" per 100 ml of the original sample.

The combination of sample volumes for processing is selected according to the type of water sample or known degree of contamination. Various configurations and tables may be used; typical volumes and dilutions are summarized in Table 4.4.

Appropriate volumes of water are added aseptically to tubes or other vessels containing sterile nutrient medium of a concentration that will ensure the mixture corresponds to single-strength medium. For example, 10 ml of sample would typically be added to 10 ml of double-strength medium or 1 ml of sample to 10 ml of single-strength medium and so on. The tube must also contain a small inverted glass tube (Durham tube) to facilitate the detection of gas production. Growth in the medium is confirmed by visible turbidity and/or a colour change. Tubes are incubated without resuscitation, and the number of positive reactions is recorded after 24 and/or 48 hours, depending on the type of analysis.

### *Presence–absence tests*

Presence–absence tests may be appropriate for monitoring good-quality drinking-water where positive results are known to be rare. They are not quantitative and, as their name suggests, they indicate only the presence or absence of the indicator sought. Such results are of very little use in countries or situations where contamination is common; the purpose of analysis is then to determine the degree of contamination rather than indicate whether or not contamination is present. Thus, presence–absence tests are not recommended for use in the analysis of surface waters, untreated small-community supplies, or larger water supplies that may experience occasional operational and maintenance difficulties.

### 4.2.3 Choice of methods

Very often the choice between the multiple-tube and the membrane-filtration methods will depend on national or local factors, e.g. the equipment already available or the cost of certain consumables. The advantages and disadvantages of each method should be considered when a choice has to be made; these are summarized in Table 4.5. The schematic decision-making network shown in Fig. 4.3 will aid the selection of procedure and method. The purpose of this diagram is merely to provide suggestions for the approach to be used; local or other circumstances will also affect the final decision.

### 4.2.4 Minimizing the cost of analysis

It is sometimes clear that faecal contamination exists (e.g. immediately downstream of a sewage discharge) or that contamination is very unlikely (e.g. in a distribution network with a free chlorine residual greater than 0.5 mg/litre, median turbidity less than 1 NTU, and pH less than 8.0). Microbiological analysis may then be deemed unnecessary. This is not appropriate, however, under certain conditions, e.g. where there is a legal requirement to conduct analysis, or where legal action that may be taken would depend on the results of a microbiological analysis of the water.

Omission of microbiological analysis under the appropriate conditions mentioned above may contribute to minimizing costs. It may also ensure that adequate numbers of samples are investigated overall where the resources available

***Table 4.5 Comparison of methods for analysis of coliform bacteria***

| Most probable number method | Membrane-filtration method |
|---|---|
| Slower: requires 48 hours for a negative or presumptive positive result | Quicker: quantitative results in about 18 hours |
| More labour-intensive | Less labour-intensive |
| Requires more culture medium | Requires less culture medium |
| Requires more glassware | Requires less glassware |
| More sensitive | Less sensitive |
| Result obtained indirectly by statistical approximation (low precision) | Result obtained directly by colony count (high precison) |
| Not readily adaptable for use in the field | Readily adaptable for use in the field |
| Applicable to all types of water | Not applicable to turbid waters |
| Consumables readily available in most countries | Consumables costly in many countries |
| May give better recovery of stressed or damaged organisms under some circumstances | |

### *Fig. 4.3 Decision-making network for selection of method of analysis*

*Note:* Analysis may sometimes be necessary because of specific local circumstances, e.g. where legislation demands that such analysis should be undertaken, or where legal action may be taken on the basis of analytical results.

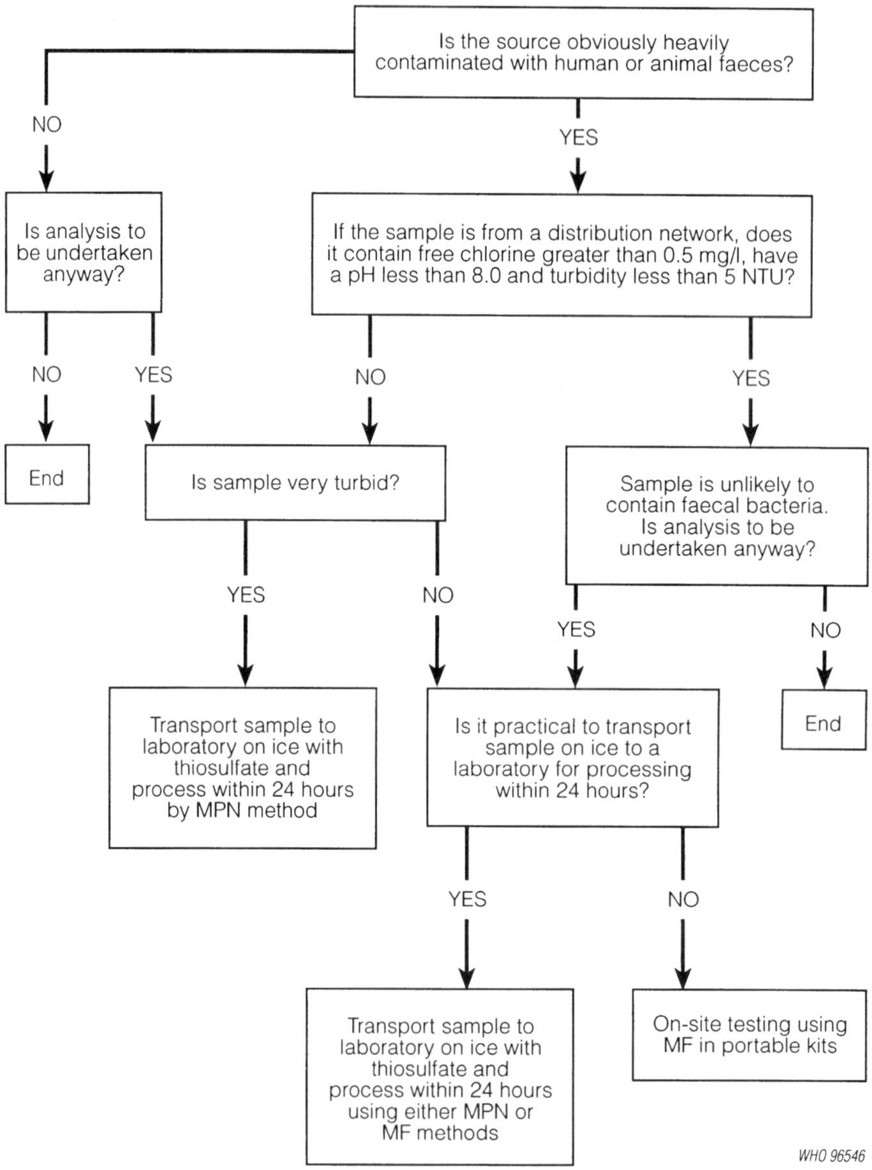

for analysis are inadequate to undertake the recommended number of microbiological analyses.

### 4.2.5 Laboratory-based versus on-site testing

Water-quality testing in communities may be subject to the following problems, especially when the communities or the sampling sites are remote or inaccessible:

- — deterioration of samples during transport to centralized laboratory facilities;
- — high cost of transporting samples;
- — inadequate techniques for sample storage and preservation during prolonged transport, thus limiting the sampling range;
- — increased personnel costs because of the need for repeat sampling journeys;
- — the need for reporting, which may necessitate further return journeys.

If there are delays in sample transport and analysis—and therefore in reporting—remedial action is also likely to be delayed. For these reasons, on-site water testing using portable equipment is appropriate in many remote areas. Portable equipment is used in many developing countries, and does help to overcome a number of logistic and financial constraints. However, it varies widely in technical specifications, including the range of analyses that can be performed, the range of methods employed, its robustness, the degree of independence from central laboratory facilities, its portability, and requirements for consumables.

Portable testing equipment may also be favoured by agencies that undertake project monitoring in more than one area on a non-routine basis and therefore prefer portability to the establishment of a conventional laboratory. For reasons that include the following, portable equipment may also be used in conventional laboratories in place of normal laboratory equipment, especially when the number of analyses to be performed per day is relatively low.

- Independence from (unreliable) power supplies. Several types of portable equipment either incorporate a rechargeable battery or may be connected to an external battery. Where energy supplies are unreliable (because of either voltage fluctuation or intermittent supply), battery operation may be advantageous.
- Cost. Comparison of the costs of the equipment required, even after allowing for that needed for back-up, may show that it is more economical to provide portable testing equipment to peripheral or decentralized laboratories than conventional laboratory equipment.
- Ease of use. Because portable equipment is often designed for use by personnel who are not fully qualified in laboratory techniques, its use is usually straightforward. However, this does not obviate the need for proper training of personnel, particularly since some portable equipment may not be accompanied by clear, well-illustrated manuals in the language of the users.

Use of portable equipment in conventional laboratories also carries a number of disadvantages, including limitations in technical specifications. Although not invariably true, the requirement for portability may mean that portable equipment is of lower precision and sensitivity than conventional equipment. Moreover, while some types of portable equipment help to reduce dependence on expensive consumables that may be difficult to obtain in many countries (e.g. by employing reusable aluminium Petri dishes, rather than dishes made of disposable plastic or fragile glass), others actually increase dependence on non-standard glassware and, particularly, consumables (such as microbiological culture media in ampoules and preweighed reagents for chemical tests). These items are invariably more expensive than ordinary laboratory consumables and may be available only from the manufacturer of the portable equipment. Independence of special consumables is of particular importance for some reagents and microbiological culture media; ready-prepared liquid media in ampoules eliminate errors in media preparation but they have only limited shelf-life. This is an especially relevant consideration in developing countries, where delays in importation, variability of demand, and problems with transport may seriously reduce the remaining shelf-life of media. Under these conditions, it is preferable to supply dehydrated media—ideally in preweighed quantities—with a relatively long shelf-life.

The use of portable testing equipment may be the result of a commitment to the decentralization of testing facilities. Whether or not this is the case, it generally means that small numbers of analyses are undertaken at a larger number of sites, which has important implications for training:

- The number of personnel carrying out analyses will be greater so that the need for training will be greater.
- The personnel who are to use the equipment (and who are therefore to be trained) will not be working in the capital city, but in relatively remote areas far from training centres.
- These personnel are less likely to have received good initial training in laboratory techniques.

Thus there is actually a greater need for training when decentralized water-quality testing is contemplated, which is in contrast to the popular perception of "simplified" portable testing equipment for which little additional training is required. Many of the benefits expected from decentralized water-quality testing and/or on-site analysis are unlikely to be realized unless adequate resources are devoted to training.

### 4.2.6 Single-application (disposable) test kits

Disposable test kits are both widely marketed and increasingly used in developed countries. Their reliability may vary widely and they should be properly assessed by a reference laboratory. In developing countries, there are other drawbacks to the use of disposable kits: unit costs, which are high in developed countries, may be still higher, and the trade-off against personnel and staff costs is thus less favourable in developing countries.

## 4.3 Physicochemical analysis

### 4.3.1 Chlorine residual

The disinfection of drinking-water supplies constitutes an important barrier against waterborne diseases. Although various disinfectants may be used, chlorine in one form or another is the principal disinfecting agent employed in small communities in most countries.

Chlorine has a number of advantages as a disinfectant, including its relative cheapness, efficacy, and ease of measurement, both in laboratories and in the field. An important additional advantage over some other disinfectants is that chlorine leaves a disinfectant residual that assists in preventing recontamination during distribution, transport, and household storage of water. The absence of a chlorine residual in the distribution system may, in certain circumstances, indicate the possibility of post-treatment contamination.

Three types of chlorine residual may be measured: *free chlorine* (the most reactive species, i.e. hypochlorous acid and the hypochlorite ion); *combined chlorine* (less reactive but more persistent species formed by the reaction of free chlorine species with organic material and ammonia); and *total chlorine* (the sum of the free and combined chlorine residuals). Free chlorine is unstable in aqueous solution, and the chlorine content of water samples may decrease rapidly, particularly at warm temperatures. Exposure to strong light or agitation will accelerate the rate of loss of free chlorine. Water samples should therefore be analysed for free chlorine immediately on sampling and not stored for later testing.

The method recommended for the analysis of chlorine residual in drinking-water employs *N,N*-diethyl-*p*-phenylenediamine, more commonly referred to as DPD. Methods in which *o*-tolidine is employed were formerly recommended, but this substance is a recognized carcinogen, and the method is inaccurate and should not be used. Analysis using starch–potassium iodide is not specific for free chlorine, but measures directly the total of free and combined chlorine; the method is not recommended except in countries where it is impossible to obtain or prepare DPD.

Procedures for the determination of free chlorine residual are described in Annex 9.

### 4.3.2 pH

It is important to measure pH at the same time as chlorine residual since the efficacy of disinfection with chlorine is highly pH-dependent: where the pH exceeds 8.0, disinfection is less effective. To check that the pH is in the optimal range for disinfection with chlorine (less than 8.0), simple tests may be conducted in the field using comparators such as that used for chlorine residual. With some chlorine comparators, it is possible to measure pH and chlorine residual simultaneously. Alternatively, portable pH electrodes and meters are available. If these are used in the laboratory, they must be calibrated against fresh pH standards at least daily; for field use, they should be calibrated immediately before each test. Results may be inaccurate if the water has a low buffering capacity.

Procedures for measuring pH using a comparator are described in Annex 10.

### 4.3.3 Turbidity

Turbidity is important because it affects both the acceptability of water to consumers, and the selection and efficiency of treatment processes, particularly the efficiency of disinfection with chlorine since it exerts a chlorine demand and protects microorganisms and may also stimulate the growth of bacteria.

In all processes in which disinfection is used, the turbidity must always be low—preferably below 1 NTU or JTU (these units are interchangeable in practice). It is recommended that, for water to be disinfected, the turbidity should be consistently less than 5 NTU or JTU and ideally have a median value of less than 1 NTU.

Turbidity may change during sample transit and storage, and should therefore be measured on site at the time of sampling. This can be done by means of electronic meters (which are essential for the measurement of turbidities below 5 NTU). For the monitoring of small-community water supplies, however, high sensitivity is not essential, and visual methods that employ extinction and are capable of measuring turbidities of 5 NTU and above are adequate. These rely on robust, low-cost equipment that does not require batteries and is readily transportable in the field, and are therefore generally preferred.

Procedures for measuring turbidity in the field using a simple "turbidity tube" are described in Annex 10.

## 4.4 Aesthetic parameters

Aesthetic parameters are those detectable by the senses, namely turbidity, colour, taste, and odour. They are important in monitoring community water supplies because they may cause the water supply to be rejected and alternative (possibly poorer-quality) sources to be adopted, and they are simple and inexpensive to monitor qualitatively in the field.

### 4.4.1 Colour

Colour in drinking-water may be due to the presence of coloured organic matter, e.g. humic substances, metals such as iron and manganese, or highly coloured industrial wastes. Drinking-water should be colourless. For the purposes of surveillance of community water supplies, it is useful simply to note the presence or absence of observable colour at the time of sampling. Changes in the colour of water and the appearance of new colours serve as indicators that further investigation is needed.

### 4.4.2 Taste and odour

Odours in water are caused mainly by the presence of organic substances. Some odours are indicative of increased biological activity, others may result from industrial pollution. Sanitary inspections should always include the investigation of possible or existing sources of odour, and attempts should always be made to correct an odour problem. Taste problems (which are sometimes grouped with odour problems) usually account for the largest single category of consumer complaints.

Generally, the taste buds in the oral cavity detect the inorganic compounds of metals such as magnesium, calcium, sodium, copper, iron, and zinc. As water should be free of objectionable taste and odour, it should not be offensive to the majority of the consumers. If the sampling officer has reason to suspect the presence of harmful contaminants in the supply, it is advisable to avoid direct tasting and swallowing of the water. Under these circumstances, a sample should be taken for investigation to a central laboratory.

## 4.5 Other analyses of relevance to health

Although the great majority of quality problems with community drinking-water are related to faecal contamination, a significant number of serious problems may occur as a result of chemical contamination from a variety of natural and man-made sources. In order to establish whether such problems exist, chemical analyses must be undertaken. However, it would be extremely costly to undertake the determination of a wide range of parameters on a regular basis, particularly in the case of supplies that serve small numbers of people. Fortunately, such parameters tend be less variable in source waters than faecal contamination, so that alternative strategies can be employed.

The range of health-related parameters may include:

- — fluoride (where it is known to occur naturally)
- — nitrate (where intensification of farming has led to elevated levels in groundwater)
- — lead (in areas where it has been used in plumbing)
- — chromium (e.g. in areas where it is mined)
- — arsenic (in areas where it is known to occur naturally)
- — pesticides (where local practices and use indicate that high levels are likely).

If these or any other chemicals of health significance are thought to be present, they should be monitored and the results examined in the light of the WHO guideline values and any relevant national standards (see Volumes 1 and 2).

Some health-related parameters may be measured in community supplies by means of portable test kits based on conventional titrations, comparators, or photometers. If this is done, the reagents must be of high quality and carefully standardized. Other parameters require conventional laboratory analysis by spectrophotometry, atomic absorption spectroscopy, or chromatography, using standard methods.

## 4.6 Analytical quality assurance and quality control

Standard methods for drinking-water analysis should be tested under local conditions for accuracy and precision, agreed at national level, and applied universally by both water-supply and regulatory agencies. However, the use of standard methods does not in itself ensure that reliable and accurate results will be obtained.

In the context of analytical work, the terms quality assurance and quality control are often treated as synonymous. In fact, they are different concepts.

Analytical quality control is the generation of data for the purpose of assessing and monitoring how good an analytical method is and how well it is operating. This is normally described in terms of within-day and day-to-day precision.

Analytical quality assurance, by contrast, comprises all the steps taken by a laboratory to assure those who receive the data that the laboratory is producing valid results. Quality assurance thus encompasses analytical quality control but also includes many other aspects such as proving that the individuals who carried out an analysis were competent to do so, and ensuring that the laboratory has established and documented analytical methods, equipment calibration procedures, management lines of responsibility, systems for data retrieval, sample-handling procedures and so on.

A checklist for effective analytical quality assurance is given in Table 4.6.

Quality assurance as applied to conventional laboratories is relatively straightforward. It is also important in field testing in view of the more exacting conditions under which it takes place and the unspecialized nature of the responsible staff. Paradoxically, therefore, quality assurance has the greatest importance in circumstances where it is most difficult to undertake. The following are among the possible approaches to overcoming the problem:

- Supervision. An effective network for on-site testing cannot function without adequate supervision, which should cover all field activities, including water-quality testing. This helps to maintain adequate standards of analysis.
- Blank sample analysis. It is unlikely that staff will be willing to submit reports from the field which question their own ability. Furthermore, it is often impractical to prepare, distribute, and collect the results of known quality-

**Table 4.6 Checklist for effective analytical quality assurance**

**Do laboratory personnel have:**
- clearly defined responsibilities?
- qualifications?
- experience?
- training?

**Is space:**
- adequate for the types and number of analyses being undertaken?

**Is equipment:**
- adequate?
- regularly serviced and maintained?
- calibrated and used only by authorized personnel?

**Are materials:**
- bought from a reliable supplier, who carries out quality control?

**Are there proper facilities:**
- for the receipt and storage of samples, and systems for coding and identifying them?

**Are data:**
- archived?
- retrievable?

**Are methods:**
- validated?
- documented?
- monitored (i.e. the results subjected to analytical quality control)?

**Is safety assured by:**
- adequate working and waste-disposal procedures?
- training of staff?
- proper maintenance of equipment?
- proper supervision of staff?

control samples, which would anyway receive especially careful treatment in the field. An alternative strategy is therefore to encourage staff to process sterile distilled water in place of the sample from time to time. If contamination does occur, analysts should then recognize the inadequacies in their own technique and question their own work accordingly. Similarly, samples known to be contaminated may be processed to provide a crude positive control.

- Equipment review. A commitment to decentralized testing with field test kits and other portable equipment normally results in a larger quantity of equipment being in use. Regular review of the equipment (e.g. temperature checking of incubators) is essential. To ensure standardization, this should be undertaken by supervisory staff from a control laboratory.

The applicability of methods under field conditions should be assessed by a central laboratory.

## 4.7 Safety

The safety of staff undertaking analytical procedures, both in the field and in the laboratory, is of the greatest importance. All staff should be trained in safety procedures relevant to their work. In the laboratory, individual staff members should be authorized to undertake procedures involving risk of any type only after appropriate training; unauthorized staff should not be allowed to undertake analyses.

All laboratories should formulate and implement a safety policy that should cover cleaning, disinfection, and the containment of hazardous substances. Safety equipment such as fire extinguishers, safety glasses, and first-aid kits should be suitably located, and readily available; they should be routinely checked and all staff should be trained in their use.

# 5. Data analysis and interpretation

## 5.1 Introduction

The objective of surveillance is not simply to collect and collate information, but also to contribute to the protection of public health by promoting the improvement of water supply with respect to quality, quantity, coverage, cost and continuity.

Clearly, the aim of a surveillance programme is to generate data that lead to optimization of activities and investment and thence to improved drinking-water supplies. Data analysis and interpretation are therefore fundamental components of the surveillance process.

## 5.2 Results of community surveys

### 5.2.1 Evaluation of water-supply systems

As outlined in Chapter 1, the evaluation of community water supplies requires the consideration of a number of quantitative factors. The quantitative nature of the evaluation makes possible the meaningful comparison of systems, and assists in the assignment of relative priorities to those requiring improvement. The indicators most commonly used to evaluate community water supplies are *quality*, *quantity*, *coverage*, *cost*, and *continuity*, as defined in Chapter 1. Each is discussed below in the context of the analysis and interpretation of the data generated during surveillance activities.

#### *Quality*

The target for water quality should be compliance with national standards, which should in turn be based on the health criteria given in Volume 1. Water quality is assessed by means of sanitary inspections and appropriate analytical measurements, discussed in detail in sections 5.3 and 5.4, respectively.

#### *Quantity*

Estimates of the volume of water needed for health purposes vary widely. It is assumed here that daily per capita consumption of drinking-water is approximately 2 litres, although this figure varies from country to country. However, this does not take into account the water needed for personal and domestic hygiene, which are also important for the maintenance and improvement of public health. In rural areas, daily consumption for these purposes varies widely; in urban areas, with piped supplies to house connections, it may exceed 100 litres per capita per day.

Measurements of the volume of water collected or supplied for domestic purposes may be used as a basic hygiene indicator. Some authorities use a guideline value of 50 litres per capita per day, but this is based on the assumption that personal washing and laundry are carried out in the home; where this is not the case, lower figures may be acceptable.

In the analysis of bulk figures related to water entering piped distribution systems, it should be borne in mind that:

- The figures will be averages, and consumption in different households may vary widely, e.g. with socioeconomic status.
- Leakages may make a significant contribution to apparent consumption.
- Even a single dwelling using piped water for irrigation or for commercial purposes may significantly influence the apparent consumption for a community water supply.
- The flow of water entering the distribution system during the day does not necessarily represent the sustained input during 24 hours, and overflows may be significant at certain times.

### *Coverage*

From the public health standpoint, the percentage of the population provided with drinking-water—the coverage—is the most important single indicator of the overall success of a water-supply programme. From the point of view of the water-supply agency, coverage is expressed as the percentage of the total population served; it may be subdivided into the population served by domestic connections, by public standposts, and by point sources such as wells and springs.

However, the surveillance agency has a responsibility for the public health aspects of water supply to the entire population. It is therefore essential that the agency undertake wider surveys of the various means by which drinking-water is provided to the population, the estimated population served by each means of supply, and the relative health risk associated with each of them. This information should be formally reported to the national planning authorities and used to guide water-supply programmes and funding strategies.

### *Cost*

Cost may be an important factor influencing access to water, and is especially important in periurban areas where water is purchased from vendors. Where such water is the only water available for personal and domestic hygiene purposes, the adverse effects of high costs on public health are proportionally greater. In these circumstances it is quite common for the amount paid by individual families for water to be sufficient, if combined, to finance the construction or expansion of a piped water supply adequate to satisfy public health needs. Information on the cost per family is therefore important for national and regional planning purposes.

Cost is also important in community water supplies where the local capacity to finance operation and maintenance is limited, especially if inappropriate technology has been employed. Where the surveillance agency identifies problems of this type, it is vital that the national and regional planning structures are informed, so that the situation will not be repeated and adequate support for operation and maintenance is provided.

Cost recovery is essential if a water supply is to be sustainable; it requires a rational charging structure. Charges must be collected and used for the purpose intended. Consumers are reluctant to pay for a poor-quality service, and this may compound the problem. Various forms of cost recovery are used, including metering, flat rates for domestic use, and charges related to the size or value of properties. Metering is often favoured, but may meet resistance from consumers; it can be costly in both installation and subsequent reading and charging.

### Continuity

Analysis of data on continuity of supply requires the consideration of two components—daily and seasonal continuity. Continuity can be classified as follows:

- year-round services from a reliable source with no interruption of flow at the tap;
- year-round service with daily variation, of which the most common causes are:
  - restricted pumping regimes in pumped systems, whether planned or due to power failure;
  - peak demand exceeding the flow capacity of the conduction line or the capacity of the reservoir;
- seasonal service variation resulting from source fluctuation, which typically has three causes:
  - natural variation in source volume during the year
  - volume limitation because of competition with other uses such as irrigation
  - periods of high turbidity when the source water may be untreatable;
- compounded daily and annual discontinuity.

This classification reflects broad categories of continuity, which are likely to affect hygiene in different ways. Thus daily discontinuity results in low supply pressure and a consequent risk of in-pipe recontamination, which is potentially hazardous in the case of unchlorinated community water supplies. Other consequences include reduced availability and lower volume use, which adversely affect washing habits. Household water storage may be necessary, and this may lead to an increase in the risk of contamination during such storage and associated handling. Seasonal discontinuity often forces users to obtain water from inferior and distant sources. As a consequence, in addition to the obvious reduction in quality and quantity, time is lost in making regular collections.

### 5.2.2 Hygiene practices

Some of the information generated by surveillance will be of interest in connection with hygiene education (see Chapter 7). Four types of information that are useful in this regard can be readily obtained:

- *Areas where hygiene education is most needed*—these may be where water is of poor quality, or where continuity is poor with the result that household storage becomes necessary.
- *The facilities available for hygiene education*—the existence of a school, community organizations, health post, or other community centre may serve to facilitate the work of hygiene educators.
- *Information on behaviour*—this can easily be collected by simple observation; observation of household water storage practices, for example, may show that water is stored in open or closed containers and is withdrawn by scooping it out by hand, by means of any available container or a container reserved for the purpose, or by means of a tap or syphon.
- *Information on the preferred means of communication*—this should cover radio and television, and the stations received, with a view to their use for educational programmes.

## 5.3 Assessment of the sanitary situation

Sanitary inspection forms (see Annex 2) are needed to collect information regarding specific points of risk to the water supply. This information may be used in various ways to facilitate the improvement of community water supplies. Key questions include:

- How can the data be expressed in terms of relative risk in order to compare a number of systems, including those of highest priority, and identify simple remedial measures that can be undertaken at local level?
- How many false positives, i.e. falsely identified risk points, can be tolerated without invalidating the system? In other words, is the system robust?
- How can a scoring system be developed which is sufficiently discriminatory to identify systems requiring urgent attention without overwhelming the workforce with the sheer amount of remedial action required? (There is, for

**Table 5.1 Examples of sanitary inspection risk[a] scores**

| Risk score | Risk |
|---|---|
| 0 | No observed risk |
| 1–3 | Low risk |
| 4–6 | Intermediate risk |
| 7–10 | High risk |

[a] The term "risk" as used here indicates potential danger to human health from a water source or supply. In Volumes 1 and 2, "risk" has a more precise quantitative connotation.

example, little advantage in a strategy that classifies 80% of systems as being at "very high risk" unless massive resources are available for remedial action.)

- How can the most important source(s) of pollution be identified among the number of potential sources that may have been noted?
- How can recurrent problems be identified which should be remedied by changes in strategy at national level rather than by repeated local remedial action?

For each type of water source the proportion or percentage of points recorded as positive for risk during the sanitary inspection gives a sanitary risk score. These scores can then be arbitrarily associated with different levels of relative risk (see Table 5.1).

The scores associated with various levels of risk should be selected in the light of local circumstances. Because the objective is to produce a classification that facilitates remedial action, it is important to ensure that the proportion of supplies or point sources falling into each category is reasonably balanced. In the early stages of implementation a narrow range of scores in the "high-risk" category may be advisable in order to avoid overloading the workforce.

It is a relatively simple matter to grade point-source systems where there are typically only 10 points for inspection, but more complicated to grade community water-supply systems which sometimes include a number of sources, treatment plants, and reservoirs, plus a distribution system. In the latter case it is particularly important to rely not only on numerical comparisons obtained by analysis of sanitary inspection data but also on an understanding of the overall functioning of the water supply. This highlights the importance of adequate training related to the water-supply practices in the locality or region concerned.

## 5.4 Microbiological water quality

As with sanitary inspection, data on microbiological water quality may usefully be divided into a number of categories; the levels of contamination associated with each category should be selected in the light of local circumstances. A typical

***Table 5.2 Example of classification and colour-code scheme for thermotolerant (faecal) coliforms or E. coli in water supplies***

| Count per 100 ml | Category and colour code | Remarks |
|---|---|---|
| 0 | A (blue) | In conformity with WHO guidelines |
| 1–10 | B (green) | Low risk |
| 10–100 | C (yellow) | Intermediate risk |
| 100–1000 | D (orange) | High risk |
| >1000 | E (red) | Very high risk |

classification scheme is presented in Table 5.2, based on increasing orders of magnitude of faecal contamination.

Where community water supplies are unchlorinated, they will inevitably contain large numbers of total coliform bacteria, which may be of limited sanitary significance. It is therefore recommended that the bacteriological classification scheme should be based on thermotolerant (faecal) coliform bacteria or *E. coli.*

Grouping of point sources into categories of the type shown in Table 5.2 is generally straightforward. Occasionally, however, where a number of samples are taken each year, the levels of faecal contamination may vary widely between successive samples. The reasons for this are often obvious and may be related to seasonal influences such as rainfall.

However, where piped small-community water supplies are being analysed and samples are taken at various points in the system, water quality may differ in different parts of the system at any one time. Again, the reasons for this may become obvious during the sanitary inspection or—if these differences are the result of cross-contamination or contamination caused by leaks in pipework—after resampling.

It is common to use 95% compliance criteria when assessing the results of microbiological analysis. This procedure is appropriate only where adequate numbers of samples are analysed for statistical purposes and is not generally applicable to small-community water supplies.

## 5.5 Risk assessment

For the purposes of risk analysis, the results of *E. coli* counts and sanitary inspection are combined.

Examination of the faecal grading together with the sanitary inspection risk scores for a large number of facilities should make it possible to assess relative priorities both for local remedial action and for regional planning purposes. In general, the classification schemes shown in Tables 5.1 and 5.2 facilitate such risk analysis when combined as illustrated in Fig. 5.1. Nevertheless, it may be necessary to test various classifications to find the combination most useful for local conditions.

***Fig. 5.1 Example of assessment of priority of remedial actions by risk analysis***

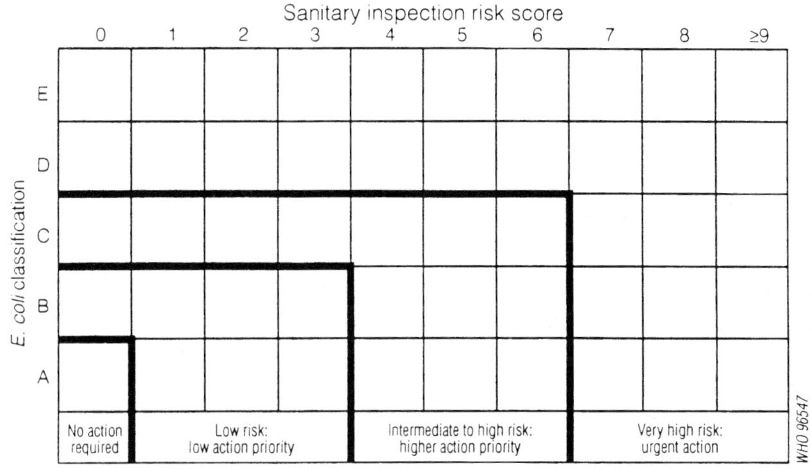


***Fig. 5.2 Example of a completed risk analysis***

*Note:* Each number represents a water-supply facility.

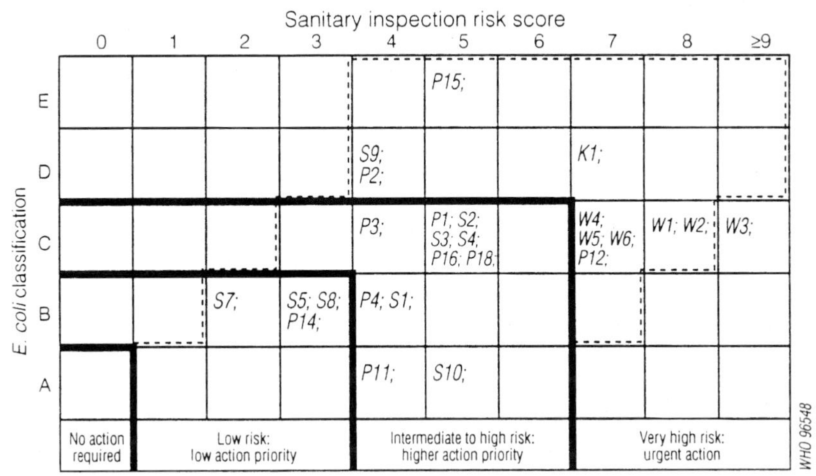


Figure 5.2 illustrates the use of risk analysis in practice. It is clear that there is a general tendency for the results to be distributed in a band running from the top right to the bottom left of the table. This is to be expected since a greater risk of contamination is likely to be associated with the occurrence of a greater degree of contamination. Nevertheless, a high sanitary risk score associated with low-level faecal contamination still requires urgent action, as does a low sanitary risk

score associated with high-level faecal contamination. It can be seen that the priority rating of such systems is high.

It is worth emphasizing that the analysis is representative of only one moment in time, whereas the inspection takes account of the previous history of the installation and future points of risk. It is this that makes the combined analysis useful—and particularly important when surveillance staff are dependent on a single bacteriological analysis or a number of analyses undertaken on a single date.

## 5.6 Presentation of information

Information must be presented in a form that is intelligible and useful to the recipients. The means of presenting the results obtained by monitoring community water supplies merits particular attention as there are several target audiences, each with different information needs and different perceptions and understanding of water-supply issues.

### 5.6.1 Target audiences

In general, the target audiences are likely to include *local system operators, community members or their representatives* with limited knowledge of matters such as water quality. For this group, it is recommended that data should be interpreted in the light of national standards or goals rather than presented only in "raw" form. Reporting to the community should generally take place as soon as possible after monitoring is completed. Where remedial actions are shown by sanitary inspection to be necessary, these should be noted. If possible, information should be provided about remedial actions that are possible at local level and those that require external assistance. Sources of information and/or advice for the former and sources of support for the latter may be noted. Where appropriate, the report may also be sent to agencies who would be responsible for providing the external assistance. The possibility of using a pictorial form which may also serve as the sanitary inspection form has been noted earlier (see section 3.3.2), examples are given in Annex 2. In general, presenting data in an easily understandable form, e.g. pictorially or by means of colour coding, is more likely to result in action by personnel at the local level.

Where a situation merits action at the individual or household level (e.g. when the severity of contamination indicates the need for household treatment), information should be disseminated not only to the community but also to the individuals and/or households concerned. "Multipliers" such as schools, clubs, and community meetings may be used for this purpose.

*Regional planners and engineers* whose responsibilities or areas of influence cover a number of supplies form another important target audience. This group often includes external agencies—both bilateral and multilateral aid agencies and NGOs—as well as national authorities.

The information required by this group is very different from that needed by the community, and consists essentially of data useful for regional planning purposes. Most information will therefore be presented in an annual report, although more frequent reporting of high-priority actions may be required. Typically, an annual report will outline the overall quality of the water-supply service in the region and identify the systems in most urgent need of attention from a public health viewpoint. Priorities can be rated as described in section 5.5. It may also sometimes be possible to indicate the nature and extent of the work required, e.g. "replace storage tank damaged by subsidence", or "increase coverage, currently 45% of a total population of 1850".

The timing of the delivery of these reports is vital and should be such as to enable them to be used in the preparation of annual work plans and budgets. An example of the form that such a regional annual report may take is shown in Annex 11.

*National planners*, a third likely target audience, will use surveillance information for large-scale planning purposes. Information intended for this group should highlight geographical priorities and major national problems. Reporting methodology should be standardized nationally to allow reasonable comparisons to be made between regions. National reports typically resemble regional reports both in presentation and timing; an example is shown in Annex 11. In general, information at this level should be presented in a highly digested form suitable for a nontechnical audience.

### 5.6.2 Simple data presentation

Experience has shown that data presented in an appropriate, generally highly simplified, form is both educative and easy for nontechnical groups, and especially local and planning staff, to understand. Material should therefore be prepared with this in mind.

At local level, a simple classification of the performance of facilities, for instance by colour coding of the type shown in Table 5.2, tends to generate competition among communities and system operators and motivation for operation and maintenance. Experience has shown that improvement is effected without substantial external inputs, probably through more effective use of the available technical facilities. For monitoring purposes, classifications of this nature facilitate the comparison of results and thus the assessment of improvement or deterioration.

In planning at both regional and national levels, the principal uses of surveillance information include policy- and strategy-making, the estimation of resource requirements and water-resource planning, and the identification of priorities for investment. The method of data presentation should facilitate comparison of water supply (in terms of quality, quantity, coverage, cost, and continuity) in different regions, the recognition of long-term trends in these parameters at regional and national level, and the pinpointing of recurrent problems that require policy changes if they are to be overcome. The quantitative nature of the data generated should make it possible to estimate the resource requirements for trained staff for surveillance at various levels, the operational requirements for surveillance, and the investment in operation and maintenance required in water-supply improvement and expansion. Estimation of the total water-resource requirements for drinking purposes facilitates intersectoral coordination and large-scale water resource planning.

## 5.7 Use of surveillance findings

### 5.7.1 Use of data at local level

At the local level, it is especially important to ensure close collaboration between the surveillance and supply agencies. Data generated by surveillance—e.g. on quality and quantity—should be shared between these agencies to maximize their usefulness. Similarly, field staff responsible for sanitary inspection should be in close communication with the staff of the supplier (whether private, municipal, or community organization) responsible for operation and maintenance.

The information reported by the surveillance agency to the supplier at local level should therefore be both detailed and appropriate to the user (e.g. the water-supply operator). However, especially with regard to water-quality data, interpretation in the context of national legislation is essential. Furthermore, some analysis of long-term trends with respect to quality, quantity, continuity, coverage, and cost, and an overall analysis of service quality, e.g. on an annual basis, facilitates the work of both agencies in ensuring adequate resources for the water-supply sector.

### 5.7.2 Regional use of data

Strategies for regional prioritization are typically of a medium-term nature and have specific data requirements. While the management of information at national level is aimed at highlighting common or recurrent problems, the objective at regional level is to assign a degree of priority to individual interventions and to prioritize remedial actions accordingly.

It is therefore important to derive a relative measure of health risk and thus establish the priority for remedial action. While the data cannot be used on their own to determine which systems should be given immediate attention (which would also require the analysis of economic and sociocultural factors), they provide an extremely important tool for determining regional priorities. It should be a declared objective to ensure that remedial action is carried out each year on a predetermined proportion of the systems classified as high-risk.

At regional level, it is also important to monitor the improvement (or deterioration) both of individual supplies and of the supplies as a whole. In this context, simple measures, such as the mean sanitary inspection score of all systems, the proportion of systems with given degrees of faecal contamination, the mean continuity or quantity of water supplied per capita per day, and the mean tariff for domestic consumption, should be calculated yearly and changes monitored.

In many countries, a high proportion of small-community water-supply systems fail to meet quality standards. However, it should be recognized that to condemn a large number of supplies is not particularly useful and may actually be counterproductive. In such circumstances it is important that realistic goals for progressive improvement are agreed with the suppliers and subsequently implemented. At no time should the surveillance agency give up its authority to demand compliance with standards; equally, however, it should recognize that the supplier should be allowed a reasonable period in which to effect improvements in the supply. Where compliance with standards is impossible (because of insuperable technical difficulties or extreme budget limitations) or would be counterproductive (because it would divert resources from other improvements of greater public health importance), the surveillance agency may elect to postpone action until the situation improves.

### 5.7.3 Use of data for national planning

At national level, priorities should be set and disseminated by means of an annual report with recommendations. The circulation list for this report should include all surveillance and supply agencies, the national planning authorities, and agencies involved in coordination within the water-supply sector, e.g. government ministries responsible for local government, natural resources, health, and finance, and external support agencies. Information exchange with the national planning authorities may provide a basis for a mutually supportive relationship between the surveillance and supply agencies.

To promote prioritization of remedial measures at national level, it is most important that information flow to the national centre is efficient, that all information generated is received, and that the national centre has the means with which to undertake the analysis of this information.

Setting priorities at national level is by its very nature a long-term process and there is often therefore little urgent need for data. Provided that specific information on individual water supplies is available from the regional centres rapidly on request, it is not necessary for the national centre to receive frequent updates for its database; periodic updates may be adequate.

# 6. Technical interventions

## 6.1 Prevention and remedial measures

Surveillance is the process of gathering systematic information on hazards in water supplies. It enables appropriate preventive measures to be taken before failure or contamination occurs. Quality control and sanitary surveys are integral parts of surveillance which, for most community supplies, is still a medium-to-long-term undertaking. Surveillance planners and coordinators must look beyond the day-to-day problems and begin to develop infrastructures and policies that address the causes of water-supply failure and contamination.

Remedial measures include all those technical and social interventions designed to improve the water-supply service. This chapter deals with interventions of a technical nature, while social issues are addressed in Chapter 7. Interventions to improve water-supply service should include community education and management training; advising on all types of remedial action, not just technical interventions, is a key role of the surveillance agency.

Economic analysis shows that it is more cost-effective to carry out regular and diligent preventive maintenance than simply to operate equipment until it breaks down and needs expensive repairs. For example, a pump that undergoes regular maintenance such as greasing and tightening of nuts will last longer and perform better than one that is not maintained, breaks down, and requires spare parts. The cost of spare parts and skilled labour is always greater than that of a pot of grease. Complete breakdowns in supply lead to reductions in water availability and sometimes also in quality, which jeopardize the health of the community.

In some countries, preventive maintenance can only be really effective if the community is also involved. However, this does not mean that governments should abrogate their responsibilities for providing support to communities that take on the burden of maintenance. A systematic approach to maintenance is needed, taking account of environmental conditions, local culture, affordability and user involvement. For example, as a general rule, the cheaper and simpler the equipment the less maintenance it requires, the more reliable it is in practice, and the easier it is to repair. Apart from the choice of equipment, other factors that need to be considered collaboratively in choosing a maintenance system include institutional responsibilities and legal obligations, logistics, financial viability, manpower training and support, and monitoring and control. The involvement of users in decision-making with regard to level of service, type of equipment, and operational system is essential to successful maintenance. Advising on the types and suitability of the remedial action to be taken should be the responsibility of the surveillance agency.

Even with adequate maintenance, surveillance and quality control will at times reveal the need for corrective action. Some sanitary deficiencies may be easy to remedy, and it may be well within the capacity of the community to take the necessary action; others may require measures that would be costly or difficult for the community to take without external financial or technical support. It is the responsibility of the sanitary inspector to correctly determine the most appropriate body to take remedial action and the urgency with which it should be undertaken. The relative urgency of some typical preventive and remedial measures is shown in Table 6.1.

Where water quality is so poor that there is an immediate threat to public health, it may be necessary to recommend emergency precautions such as boiling drinking-water or to supply chlorine tablets for disinfection at household level. The water supplier or surveillance agency should ensure that remedial measures are promptly executed, and then carry out a bacteriological analysis of the supply to determine whether it is safe to use.

Water-supply agencies should systematically evaluate maintenance practices in order to pinpoint difficulties and find the most effective maintenance system. An overview of the principal maintenance requirements of different types of water-supply system is necessary to assist in the selection of equipment. Where users are directly responsible for their water supplies, there should be an adequate community-based management system based on local organizational structures and integrated into the institutional hierarchy of the water-supply agency.

## 6.2 Protecting water sources

If water supplies are to remain potable, both the source and the catchment need protection. A watershed that is used to supply untreated surface water should be sparsely inhabited and should consistently yield clean, clear water. Every effort should be made to site the abstraction point above sources of pollution; if this is not possible, appropriate forms of treatment must be applied (see section 6.6). An example of a sanitary inspection form for a simple, preliminary type of sanitary inspection of surface-water abstraction is given in Annex 2.

### 6.2.1 Catchment protection

A survey of the catchment area should reveal potential sources of contamination. Surface waters and groundwaters are both vulnerable. Whereas raw-water reservoirs may be protected from large-scale human activity, rivers may pass through heavily populated areas and be contaminated by both domestic and industrial discharges. Groundwaters may be contaminated by the seepage of industrial

***Table 6.1 Preventive and remedial measures***

| Source and mode of supply | Evidence or information available | Immediate remedial measures | Preventive action for avoiding recurrence |
|---|---|---|---|
| Untreated community rainwater collection systems | Localized epidemic of enteric infection | Chlorinate water in collection reservoir (tank, container, etc.) or recommend boiling or disinfection in the home | (a) Ensure that collection surfaces are in a sanitary condition and that bypass for initial collected water is properly operated<br>(b) Promote community education and participation |
| Open dug wells | Findings of sanitary inspection unsatisfactory<br><br>Localized epidemic of enteric infection | (a) Clean well if necessary and shock-chlorinate, followed by continuous chlorination<br>(b) Recommend boiling of drinking-water, use of disinfectants and/or filters in the home | Convert to a protected, covered well with hand-pump or device for raising water isolated from the user; discourage construction of new open dug wells; promote community education and participation |
| Unpiped supplies from covered wells or shallow or deep tubewells with hand pumps or motorized pumps | Findings of sanitary inspection unsatisfactory | Confirm bacteriological quality and if necessary recommend boiling or use of disinfectant and/or filters in the home | Eliminate pollution sources and/or repair well if necessary to remedy shortcomings found in sanitary inspection. |
| | Localized epidemic of enteric infection | (a) If an alternative safe supply is not available, recommend boiling or use of disinfectants in the home<br>(b) Confirm bacterial quality<br>(c) Conduct a detailed sanitary inspection and remedy shortcomings found | (a) Take opportunity to promote community education and participation<br>(b) Feed information on the episode and sanitary survey results back to the water-supply agencies to help in deciding whether the technologies used and the codes of practice are appropriate |

| | | | |
|---|---|---|---|
| Untreated piped supplies | Findings of sanitary inspection unsatisfactory | Confirm bacteriological quality and if necessary recommend boiling or use of disinfectant and/or filters in the home | Eliminate pollution sources and/or repair system if necessary to remedy shortcomings found in sanitary inspection |
| | Unsatisfactory bacteriological quality of water at source | (a) Chlorinate supply if feasible or recommend boiling or disinfection in the home<br>(b) Conduct a detailed sanitary inspection and remedy shortcomings found | Protect the source and its catchment (this is very important) |
| | Unsatisfactory bacteriological quality of water in the distribution system | (a) If source is unsatisfactory, proceed as above<br>(b) If source is unsatisfactory but distribution system is suspected, chlorinate supply or recommend boiling or disinfection in the home<br>(c) Conduct a detailed inspection of distribution system and remedy shortcomings found | Frequent and improved supervision of the distribution system and prompt repair and maintenance are essential, especially for intermittently operated systems |
| | Localized epidemic of enteric infection | (a) Take sample for bacteriological quality determination; without waiting for this result, chlorinate general water supply or recommend boiling or disinfection in the home<br>(b) Conduct a detailed sanitary inspection of source and distribution system, and remedy shortcomings found | Frequent and improved supervision of the source and distribution system is necessary; careful operation and maintenance of such systems are essential, especially for intermittent systems |

**Table 6.1** (continued)

| Source and mode of supply | Evidence or information available | Immediate remedial measures | Preventive action for avoiding recurrence |
|---|---|---|---|
| Treated piped supervision supplies | Findings of sanitary inspection of source, treatment plant, and/or distribution system unsatisfactory | Confirm bacteriological quality and if necessary recommend boiling or use of disinfectant and/or filters in the home | (a) Frequent and improved of the whole system is necessary; careful operation and maintenance are essential for intermittent systems<br>(b) Ensure that routine sanitary inspections are carried out<br>(c) Feed information back to the water-supply agencies |
| | Unsatisfactory bacteriological quality of water after treatment or in the distribution system | (a) Ensure adequate chlorination of general supply or recommend boiling or disinfection in the home<br>(b) Conduct a detailed sanitary inspection of the whole system and remedy shortcomings found | (a) Frequent and improved supervision of the whole system is necessary; careful operation and maintenance of such systems are essential, especially for intermittent systems<br>(b) Ensure that routine sanitary inspections are carried out<br>(c) Feed information back to the water-supply agencies |
| | Localized epidemic of enteric infection | (a) Take sample for bacteriological quality determination; without waiting for this result, chlorinate general supply or recommend boiling or disinfection in the home<br>(b) Conduct a detailed sanitary inspection of source and distribution system, and remedy shortcomings found | (a) Frequent and improved supervision of the whole system is necessary; careful operation and maintenance of such systems are essential, especially for intermittent systems<br>(b) Ensure that routine sanitary inspections are carried out<br>(c) Feed information back to the water-supply agencies |

wastes buried in the ground or in abandoned wells, and by chemicals discharged accidentally onto the land. Both surface waters and groundwaters are at risk from agricultural pollution in rural areas.

Where possible, protection zones should be clearly demarcated, and activities that may affect water quality should be restricted or prohibited within their boundaries. Such activities may include the dumping of toxic waste, the discharge of undesirable effluents, drilling, mining, quarrying, and the use of agricultural fertilizers and pesticides. Where restrictions are imposed, it is important to publicize the conditions under which normal activities, e.g. housing developments, farming, mining and manufacturing, are permitted within protection zones.

In some parts of the world, risk assessment of water sources and catchment areas is based on systems that take into consideration the hydrogeology, and the hydraulic loading of contaminants at and below the surface. Some governments are beginning to introduce legislation on groundwater protection zones under which housing, industrial and certain agricultural activities will be excluded from specified parts of catchment areas.

Water suppliers are beginning to recognize three protection zones for groundwater, as follows:

1. The area surrounding the source most at risk from contamination by pathogens. This is often the 50-day isochron (the area within which pathogens would reach the source in 50 days or less).
2. The area surrounding the source most at risk from chemical contamination. This will vary greatly and will depend on aquifer type and abstraction rate as well as on industrial and agricultural activity in the area.
3. The total catchment area.

The establishment of protection zones requires intersectoral agreements involving various authorities and ministries such as those concerned with health (surveillance), agriculture, forestry, housing, and environmental protection, as well as the water suppliers. The demarcation and acceptance of protection zones should be considered by governments of countries where groundwater accounts for a significant proportion of the water supply. For further information on the theoretical basis and practical application of groundwater protection zones, see p. 145, "Selected further reading".

## 6.2.2 Groundwater protection

The most significant risk to human health related to drinking-water quality is from microbiological—particularly faecal—contamination. Health protection thus demands that sources of microbiological contamination are located sufficiently far from drinking-water sources as to minimize or eliminate the health risk.

When abstraction from a water source for human consumption is being considered, the minimum safe distance (MSD) for all potentially polluting activities should be fixed during the planning stage. Both surface and groundwater sources of drinking-water require protection. However, groundwater in its natural state is generally of good quality, and because subsurface water movement is relatively slow, it is usually easier to control sources of contamination of groundwater than it is for surface-water sources. For community supplies, the commonest sources of microbiological contamination are on-site sanitation and sewage-treatment facilities, open wells and other open surface sources of water (e.g. borrow pits), and concentrated animal husbandry.

The MSD should be determined from the time taken by contaminants to travel from their source to the source of drinking-water. This will depend on local conditions, the most important of which are the geological and hydrogeological conditions of the area, the quantity of faecal matter likely to be discharged, and the number of existing and planned sources of contamination. It is therefore very difficult to specify a universally applicable minimum distance between the location of, for instance, pit latrines and a water source. In an area where the aquifer is highly permeable and the overlying unsaturated zone (see Fig. 6.1) thin and permeable, the MSD for a latrine will be far greater than in an area where a relatively thick and impermeable unsaturated zone overlies an aquifer of relatively low permeability.

***Fig. 6.1 Groundwater terminology***

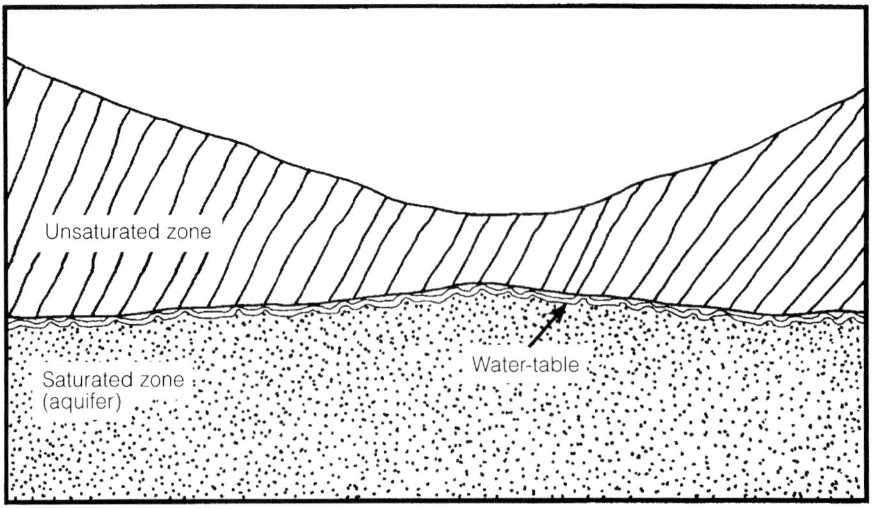


WHO 96549

In areas of fissured rock aquifers (where water is held in cracks and joints in the rock), the velocity of groundwater movement, and therefore of contaminants, will be high and must be taken into consideration when MSDs are set. This is particularly important for planning on-site sanitation where a thin, unsaturated zone of relatively low permeability overlies a fissured rock aquifer, e.g. in a karstic (weathered limestone) area. As the unsaturated zone is where the majority of microbial removal takes place, no direct source of contamination should come into contact with the water-table at its highest level.

The direction of flow of groundwater in an area will also influence the MSD. As a general rule, shallow groundwater movement reflects surface topography; sources of contamination should therefore be located downhill of drinking-water sources wherever possible.

The concentration of contaminating activities in the area concerned also affects the MSD and is particularly important where on-site sanitation or nonconventional sewage treatment is used. In areas where there are very large numbers of sources of microbiological contamination, such as low-income urban areas using on-site sanitation, there may be a build-up of nutrients in the unsaturated zone and, possibly, the aquifer. This may increase the survival time of microbes and so extend the MSD.

It is often difficult to obtain hydrogeological data in rural areas, and in community-based programmes it may not be possible to conduct thorough surveys in each area. An MSD can still be determined, however, although it may be less accurate than in other areas.

When MSDs for an area are being established, the information that will be required on the local soil and geology can be obtained by drilling or auguring to the water-table and carefully recording changes in soil and rock type, particularly changes in grain size, compaction, and the location of saturated layers. This information should then be recorded in the form of a log in which soil and rock type are plotted against depth. It is also important to carry out an infiltration test, which will give an indication of the permeability in the area. If the supply is to be a well, this can be done during test drilling (whether mechanical or by hand); where other groundwater sources, such as springs, are to be used, the infiltration test should be done in the surrounding area when the yield is tested.

Combining information from the log with data from the infiltration test will provide a good indication of the risk to the water source. Guidance on infiltration tests, infiltration rates in different types of rock, and corresponding MSDs is given in Annex 2.

Precise demarcation and enforcement of protection zones are not easy, especially where low-volume abstraction, for instance by means of hand-pumps, is practised. In these conditions, providing adequate sanitary protection of the water source and its immediate environment is likely to be easier and more effective. In much of this chapter, therefore, attention is focused mainly on the technical interventions that may be used to reduce or remove the sanitary hazards revealed by sanitary inspection on or close to the water-supply installation.

## 6.3 Wells

### 6.3.1 Dug wells

Open or poorly covered well heads pose the commonest risk to well-water quality, since the water may then be contaminated by the use of inappropriate water-lifting devices by consumers. The most serious source of pollution is contamination by human and animal waste from latrines, septic tanks, and farm manure, resulting in increased levels of microorganisms, including pathogens. Contamination of drinking-water by agrochemicals such as pesticides and nitrates is an additional and increasing problem for small-community supplies.

Dug wells are generally the worst groundwater sources in terms of faecal contamination, and bacteriological analysis serves primarily to demonstrate the intensity of contamination and hence the level of the risk to the consumer. As indicated in Annex 2, an on-site inspection can effectively reveal the most obvious sources of contamination, and can be used to promote well-head protection.

Various types of hand-dug wells are shown in Fig. 6.2, ranging from poorly protected to well protected; all types should be included by the surveillance agency in the inventory. The upgrading of unprotected wells and the construction of protected wells for community use should be strongly promoted.

Many tens of millions of families worldwide still depend on private and public dug wells; technical assessment and improvement of these wells is therefore very important. The commonest physical defects leading to faecal contamination of dug wells are associated with damage to, or lack of, a concrete plinth, and with breaks in the parapet wall and in the drainage channel. However, the most hazardous gross faecal contamination is most commonly associated with latrines sited too close to the well. Emergency relocation of either the latrines or the water source is essential when such serious problems are encountered.

An open dug well is little better than an unprotected hole in the ground if the above-mentioned physical barriers to surface-water contamination are not regularly maintained. The majority of open dug wells are contaminated, with levels of at least 100 faecal coliforms per 100 ml, unless very strict measures are taken to ensure that contamination is not introduced by the bucket. A community dug well with a windlass whereby *one* bucket is suspended over the well in a narrow opening is an improvement on each individual using his or her own bucket.

Water quality should be greatly improved by the installation of a hand-pump and the fitting of a sanitary cover to an open dug well, access being restricted by a lockable sanitary lid, which prevents any contamination of the well by buckets. However, even this relatively costly improvement may fail to reduce contamination significantly unless the well lining is made watertight down to the dry-season water-table. If faecal contamination persists, the community may have to resort to pot chlorination (see section 6.6.11), but this requires considerable organization and management to be successful; effective physical protection of the source is generally preferred.

### *Fig. 6.2 Types of hand-dug wells*

Unprotected waterhole

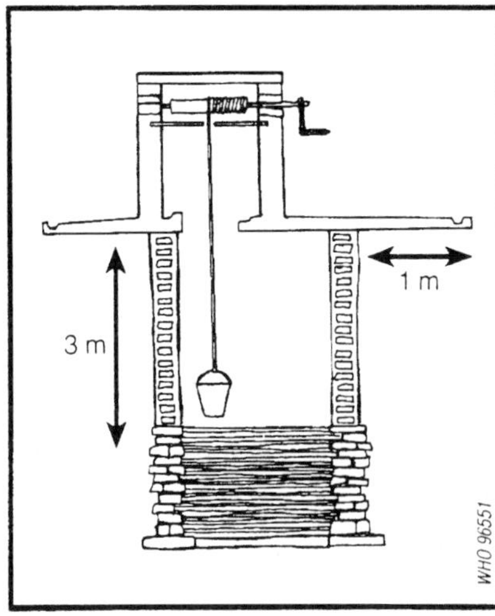


Dug well with windlass

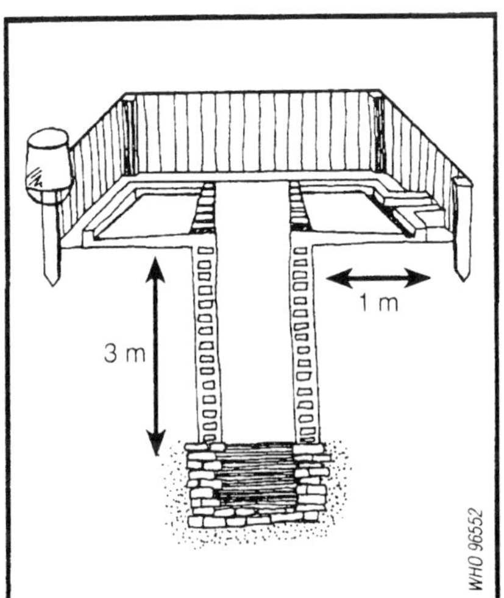


Open dug well

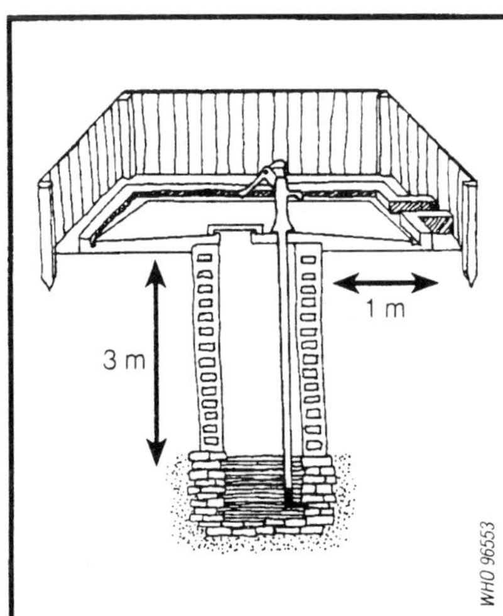


Converted hand-pumped dug well

Occasionally the aquifer itself may be contaminated; in these circumstances the only option may be to routinely disinfect the groundwater source or resort to a deeper aquifer and mechanical pumping.

### 6.3.2 Hand-pumped and mechanically pumped wells

In about 85% of cases, shallow or deep tubewells with hand-pumps and proper sanitary protection will supply water that contains few, if any, faecal indicator bacteria. Where indicator bacteria are identified, the source of faecal contamination can usually be detected by an on-site sanitary survey at and around the well-head (except where the aquifer itself is contaminated). Detailed checklists (see Annex 2) for use in inspections have been drawn up for point-source supplies in rural areas. Sanitary inspections are a useful monitoring tool and are sometimes the only affordable means of identifying water sources at risk of contamination.

To ensure that the sanitary protection of a tubewell is adequate, a reinforced concrete plinth should be built on to the well-head; its diameter should be greater than that of the riser. The plinth should be sound and drained, and the hand-pump should be located and sealed in it in a sanitary manner above the surrounding plinth and ground level. A concrete apron should be laid around the well-head and plinth, at least 2 metres in diameter and sloped towards the drainage channel, which should run to a soakaway located away from the tubewell. Additional sanitary protection should be provided by fencing the well site to keep animals out.

The area immediately surrounding the tubewell should be managed in such a manner as to reduce the risk of contamination. Latrines should be located downhill from the well and a minimum of 10 metres away from it, sources of pollution, such as open dug wells, within 15–20 metres of the tubewell should be filled in, and animals should be kept at least 10 metres away. It is difficult to define protection zones for individual tubewells as the resources are rarely available for a full study of the properties of the aquifer or for comprehensive pumping tests.

Tubewells sometimes show evidence of persistent contamination, even though sanitary inspection has revealed few local hazards. This may be the result of aquifer contamination, which is a particular problem where fissured geological strata are combined with thin top soil, and is on the increase, notably in urban and periurban areas. Under these conditions, it will be necessary either to disinfect the water supply continuously, or to locate a deeper aquifer, sink a deep borehole, and use mechanical pumping. Mechanical pumping from a deep borehole is a conventional technology more usually associated with urban settlements and developed countries because of the operation and maintenance requirements. The same principles of sanitary protection apply, and it is generally appropriate to define protection zones for the borehole because the output is much higher than that of a hand-pumped tubewell and can serve a greater population, the area of the aquifer exploited is correspondingly larger, and adequate resources are more likely to be available.

Drilling a borehole makes it possible to reach deep aquifers that are less likely to be affected by pollutants originating from the land or surface waters. Water from deep boreholes is normally free from microbiological contamination and may be used by small communities without further treatment. However, certain structural precautions are essential when wells and the associated pumps are installed. The pump casing should extend approximately 30 cm above ground and downwards to the parent rock. Concrete aprons and platforms should be constructed as for shallow wells, and the concrete sanitary seal should extend down into the space (annulus) between the casing and the excavation.

Figure 6.3 shows the sanitary protection below the pump of a deep borehole. A sanitary inspection form for this type of installation is shown in Annex 2.

### *Fig. 6.3 Sanitary protection of a deep borehole*

*Notes*: The well casing extends down to the aquifer, but the concrete sanitary seal only to a depth of 3 m. The platform (plinth) drains away from the well.

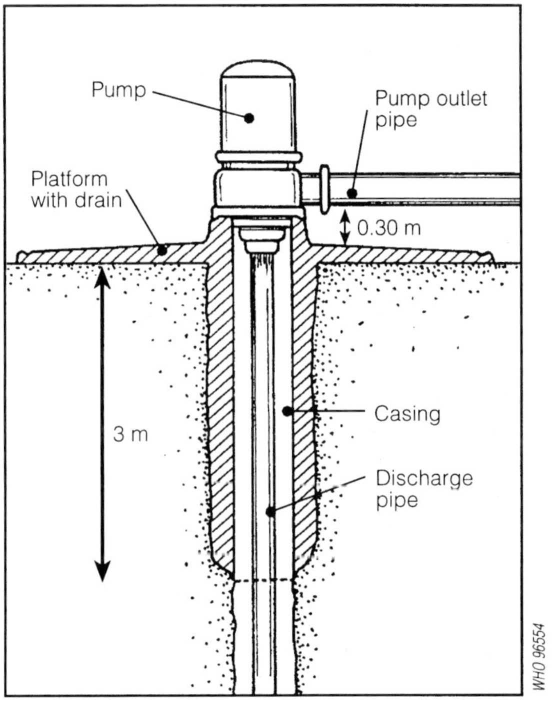

## 6.4 Springs

If a spring is to be used as a source of domestic water:

— it should be of adequate capacity to provide the required quantity and quality of water for its intended use throughout the year;

— it should be protected to preserve its quality.

A spring encasement consists of the following features (see Fig. 6.4 and Annex 2):

— spring box (watertight tank), which intercepts the source and extends downwards to an impermeable layer, or a system of collection pipes and a storage tank;

— a cover that prevents the entrance of surface drainage or debris into the storage tank;

— a protected overflow outlet;

— a connection to the distribution system or auxiliary supply;

— an impermeable layer (e.g. of concrete or puddled clay) behind the box and above the eye of the spring to prevent the infiltration of contaminants.

Provision must be made for the cleaning of the tank and the emptying of the contents.

Exposed springs are vulnerable to contamination from human and animal activities (see Figs 6.5 and 6.6). The usual method of protecting springs is to collect the water where it rises by enclosing the eye of the spring in a covered

***Fig. 6.4 Protected gravity spring***

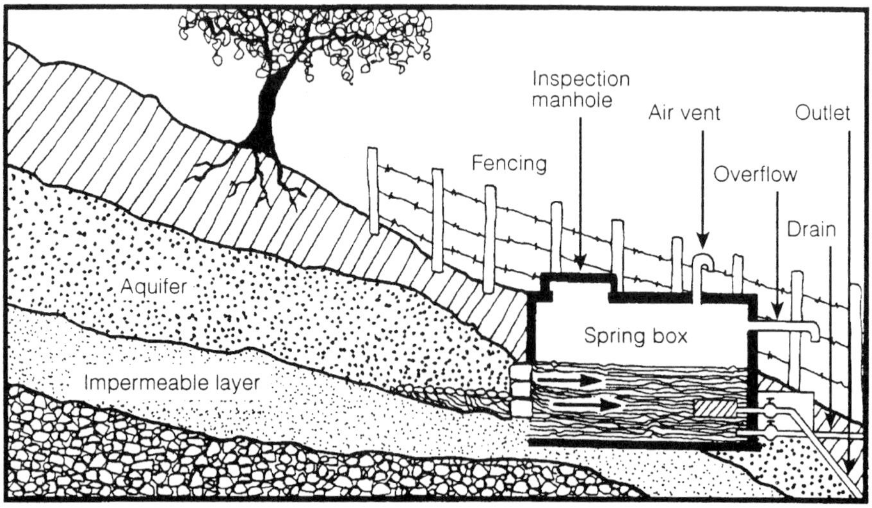


WHO 96555

***Fig. 6.5 Unprotected gravity spring***

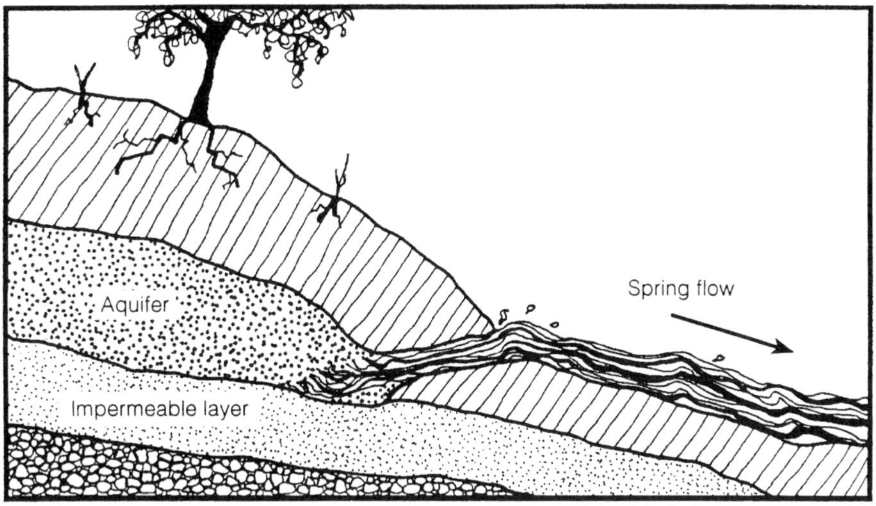


WHO 96556

***Fig. 6.6 Unprotected artesian spring***

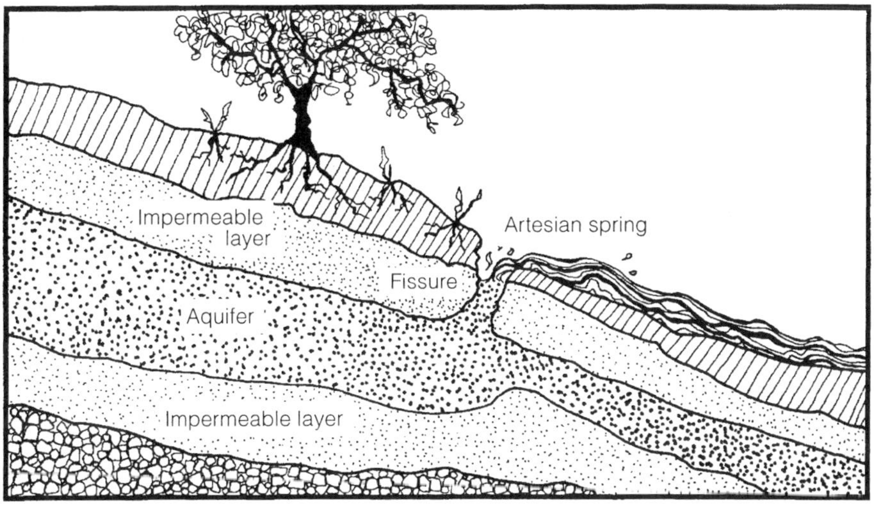


WHO 96557

chamber or box with an outlet near the bottom to allow water to flow away from the original site of the spring; in this way the natural spring is disturbed as little as possible. The exact procedure will depend on the type and site of the spring (see Figs 6.4 and 6.7). The hillside must be excavated to a sufficient depth to tap

***Fig. 6.7 Protected artesian spring***

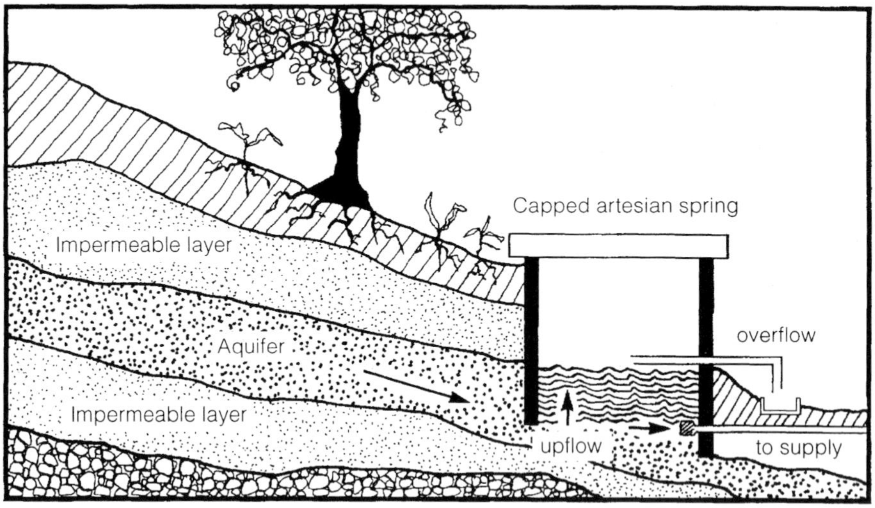


WHO 96558

the aquifer even when the water level is low and, for a protected gravity spring, to ensure that the collected water does not exert a back-pressure on the eye of the spring. The intake structure should be designed, and the excavated area backfilled with graded gravel, to prevent the inflow of sand and silt with the water into the spring box; this will form the back wall of a gravity spring and the floor of an artesian spring. The intake and gravel backfill should be covered by an impermeable cap (of concrete or puddled clay for example) to prevent surface-water infiltration. To ensure that the collected water is not contaminated, an adequate conduction pipeline and storage tank, if required, should be provided. The spring box should have a lockable inspection cover. Air vents, drains, and overflows should be fitted with mesh screens, and the whole structure surrounded by a ditch to divert surface water (Fig. 6.4). A full sanitary inspection checklist is given in Annex 2.

Springs usually become contaminated when barnyards, sewers, septic tanks, cesspools, or other sources of pollution are located on higher adjacent land. In limestone formations, however, contaminated material frequently enters the water-bearing channels through sink holes or other large openings and may be carried along with groundwater for long distances. Similarly, if material from such sources of contamination enters the tubular channels in glacial drift, water may remain contaminated even after travelling for long distances.

The following precautionary measures will help to ensure that spring water is of a consistently high quality:

- Providing for removal of surface drainage from the site. A surface drainage ditch should be located uphill from the source so as to intercept surface-water runoff and carry it away from the source. The location of the ditch and the points at which the water should be discharged are a matter of judgement, based on factors such as topography, subsurface geology, land ownership, and land use.

- Constructing a fence to prevent the entry of livestock. The location of the fence should be selected in the light of the considerations mentioned above. The fence should exclude livestock from the surface-water drainage system at all points uphill of the source.
- Providing for access to the tank for maintenance; unauthorized removal of the cover should be prevented by fitting a suitable locking device.
- Designing the cover in such a way as to prevent contamination from entering the storage tank.
- Monitoring the quality of the spring water by means of periodic checks for contamination. A marked increase in turbidity or flow immediately after a rainstorm is a good indication that surface runoff is reaching the spring.

Water from a protected spring may be supplied to small communities either directly or via a distribution system. Such systems may not be disinfected because the water is bacteriologically safe and chlorination is expensive. Where spring-fed water supplies do require disinfection, either because it is mandatory under local legislation or because of inadequate quality, this is generally done on a continuous basis: chlorine is added either as the water enters the conduction pipe from the spring box, or as it leaves a storage tank to enter the distribution network.

Artesian springs should be protected by a box with walls extending above the maximum static head; a strong sanitary cover should also be provided. To conserve water and increase the productivity of an artesian well, the casing must be sealed into the confining stratum, otherwise water may be lost through leakage into lower-pressure permeable strata at higher elevations. A flowing artesian well should be designed so that the movement of water from the aquifer can be controlled; water can be conserved if the well is equipped with a valve or shut-off device. When the recharge area and aquifer are large, and only a small number of wells penetrate the aquifer, the flowing artesian well produces a fairly steady flow of water throughout the year.

## 6.5 Rainwater catchment

Rainwater collected from clean house roofs can be of better microbiological quality than water collected from untreated household wells. When rain falls after a long dry period, however, any rainwater collected may carry with it significant amounts of contamination and debris which have accumulated on the roof and in the gutters. It is therefore recommended that the water running off the roof after the first storms of the season, and preferably for the first 5–10 minutes afterwards or until it runs clear, should be discarded or used for purposes other than drinking. Various devices are available for diverting this initial flow to waste or secondary uses.

The quality of the collected rainwater can also be improved by proper maintenance of the roof and gutters, and careful cleaning at the beginning of every wet season. Some form of mesh should be placed between the guttering and the downpipe to prevent the entry of coarse debris; it then becomes important to clean the screen regularly to prevent blockage. The worst fouling of roofs occurs when they are situated under trees in which birds roost. In areas where malaria is endemic, care should be taken to avoid creating pools of water that could become breeding sites for mosquitos.

A rainwater storage tank should be completely covered and well maintained. If the cover is inadequate, lizards and geckos will enter and produce elevated thermotolerant (faecal) coliform counts. A fine mesh fitted to all openings to the tank will prevent the entry of organic debris. Water should be drawn off by a tap located a little above the base of the tank. A sanitary inspection checklist for rainwater tanks is given in Annex 2.

## 6.6 Water treatment

For small communities, it is generally preferable to protect a groundwater source that requires little or no treatment than to treat surface water that has been exposed to faecal contamination and is usually of poor quality. In many circumstances, however, surface water is the only practicable source of supply and requires affordable treatment and disinfection. The range of treatments available for small-community supplies is necessarily limited by technical and financial considerations; the most appropriate and commonly used treatments are summarized below. Installation of packaged treatment plants is not a suitable means of dealing with the typical water-quality problems that prevail in rural areas.

### 6.6.1 Abstraction

The control measures required at the point of abstraction are determined to a large extent by the characteristics of the source water and the particular water-treatment method adopted. Screens are necessary where floating or large suspended solids are present in the source water; these will require periodic cleaning. Properly constructed intake channels or side weirs can be used to provide regular lateral intake flows from a surface-water source. Sluice-gates and valves offer a means of controlling flow but require regular maintenance and adjustment.

When upstream flow depths are controlled by properly constructed overflows, use of weirs will provide relatively fine flow control with a minimum of attention. For community supplies, the "V" notch angle required may be 45°, instead of the more common 90°, to enable a reasonable upstream depth to be achieved.

Many of the major problems of community surface-water supply begin at the point of abstraction; the following are the most common (see also Fig. 6.8 and Annex 2).

***Fig. 6.8 Surface water intake***

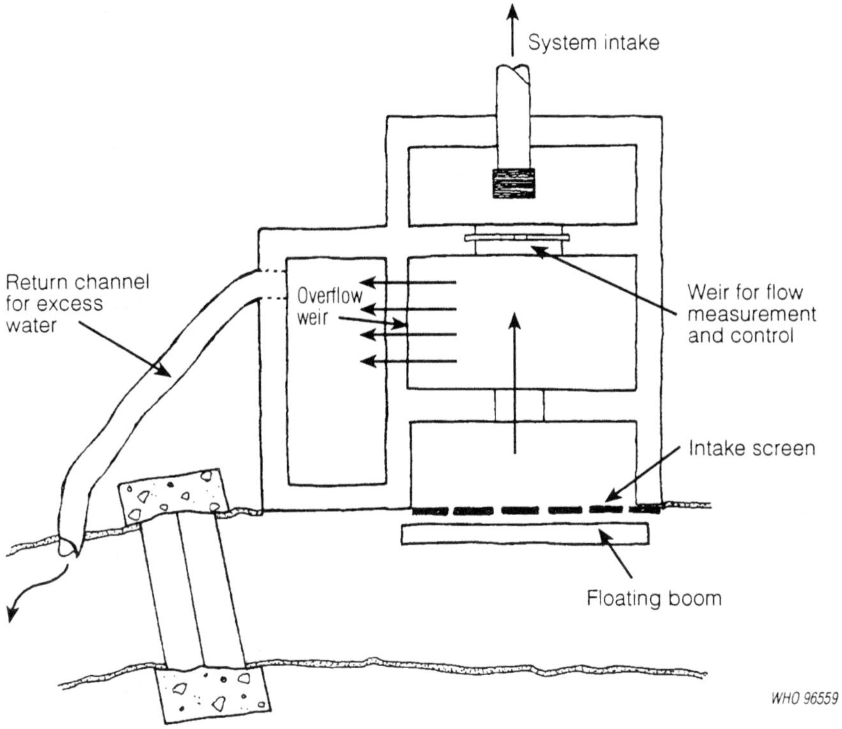


*A. Plan view*

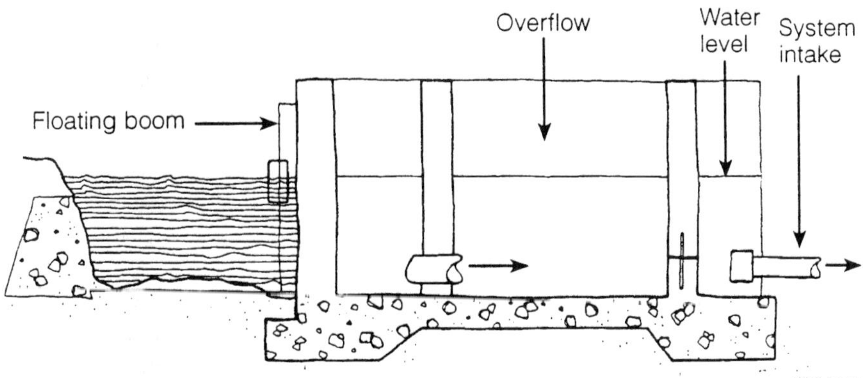


*B. Cross-section*

- There is no weir across the stream or river, and at times of low flow there is insufficient water to supply the community.
- There is no intake screen and consequently the intake is often blocked; this either causes interruptions in supply or allows large debris to pass on to the treatment plant.
- There is no floating boom at the intake, and floating substances (oils, fats) therefore pass on to the treatment plant.
- There is no flow control or the flow control is inappropriate or lacks an overflow.

### 6.6.2 Preliminary treatment by storage

Preliminary storage in a reservoir helps to guarantee a continuous supply of water despite variations in demand and in source-water availability. It can also provide an economical means of settling out some of the suspended solids.

In areas affected by schistosomiasis, protected storage for a minimum of 48 hours provides a degree of safety: the cercariae are unable to infect a host and will die. The numbers of other organisms can also be reduced in this way. If longer retention times can be achieved, the numbers of microorganisms can be significantly reduced, although this often requires storage for more than a week. However, prolonged storage in uncovered reservoirs can encourage algal growth and mosquito breeding. If the required storage volume is such that it is not practicable to construct a covered reservoir, efforts must be made to avoid the creation of habitats suitable for mosquitos, snails, or other organisms associated with disease in the surrounding communities.

### 6.6.3 Plain sedimentation

Surface waters may contain sand, grit, silt, and other suspended solids which can damage pumps, block filters, clog pipes and reduce the effectiveness of disinfection. Sedimentation helps to reduce suspended solids before treatment by filtration and can remove significant numbers of harmful organisms from polluted water. Fine silt or clay particles, however, are unlikely to be removed to any significant extent in a sedimentation tank without the use of chemical coagulation (see section 6.6.6).

Grit or coarse suspended solids can be removed in a grit tank or channel (coarse sedimentation tank) with a throughput velocity of less than 0.75 m/s and a retention time of a few minutes. The amount of finer suspended matter can be reduced by passing the water slowly through a settler or sedimenter (sedimentation tank), allowing time for it to settle out. Inlet, outlet, and internal baffle arrangements should be designed to maximize the retention time in the tank. The baffles should also assist in creating a regular flow pattern, without turbulence, throughout the tank. Construction of the sedimentation tank must be such as to permit routine desludging and cleaning operations to be carried out. The retention time in a sedimentation tank is usually significantly shorter than that for a storage reservoir, typically a few hours.

The principal problems of plain sedimenters, which can lead to poor water quality, are:

- Short-circuiting of the flow because of the absence or poor design of baffles.
- Poor maintenance, leading to the accumulation of excessive amounts of sludge and consequent carry-over. A suitable design is shown in Fig. 6.9, which also indicates the key points to be checked during sanitary inspection.

To ensure effective operation:

- The inlet baffle wall of the sedimentation tank should be perforated so that water is introduced uniformly across the entire cross-section of the tank and rapid transit across the surface of the tank is avoided.
- The floor of the sedimenter should slope towards a sludge channel, which should in turn slope towards the washout valve. It is important to ensure that:
  - the washout valve is of large diameter so that drainage is rapid;
  - the valve is functional and greased;
  - the floor of the tank is relatively clean after washout.

The effectiveness of the sedimenter should be assessed by the following means:

- Checking the turbidity at the inlet and outlet. As a guideline, an ineffective sedimenter may reduce turbidity by less than 50% but an efficient one can achieve up to 90% reduction.
- Checking the retention time. This is done by introducing sufficient salt at the inlet to increase the conductivity of a "plug" of water. The time taken for the

***Fig. 6.9 Plain sedimentation tank***

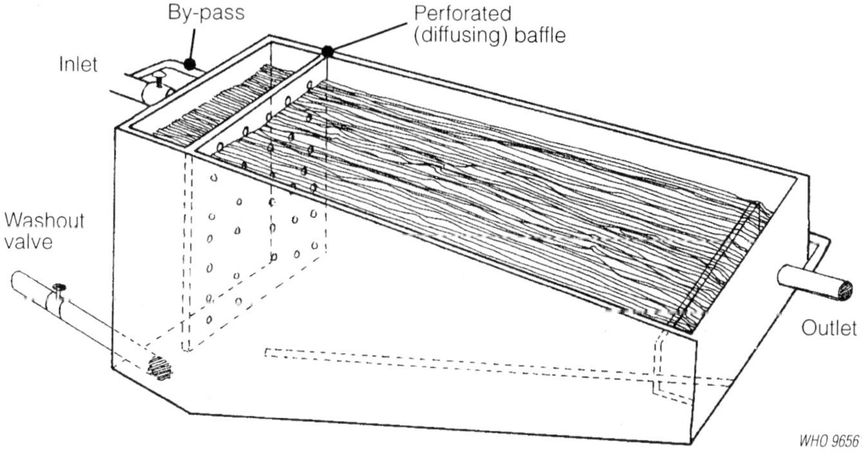

increased conductivity to be measurable at the outlet is measured, and a curve is plotted of conductivity at the outlet against time. In a well designed sedimenter, the increase in conductivity at the outlet should occur at least 2 hours after the salt was introduced at the inlet. A minimum retention time of 2 hours is essential for most types of water if removal efficiencies of greater than 50% are to be achieved.

A by-pass pipe around the sedimenter is essential to ensure continuity of flow during maintenance.

### 6.6.4 Prefiltration

In small treatment plants where the suspended solids content and turbidity of the source water are continuously or periodically high, prefiltration with gravel or other coarse material before sand filtration is an effective means of preventing the rapid blocking of the sand filters. A typical prefilter consists of a tank divided into several compartments filled sequentially with material of sizes ranging from very coarse, e.g. 50-mm pebbles, in the upstream compartment to fine, e.g. gravel 6–10 mm in diameter, in the downstream compartment. Raw water is passed vertically or horizontally through the different compartments and is then collected in an outlet chamber. If vertical flow is chosen, either upflow or downflow is possible, but upflow filters are easier to clean and thus more likely to operate effectively.

Typical filtration rates for three-stage gravel prefilters are in the range 0.5–1 $m^3/m^2$ per hour. The lower loading is appropriate for raw waters of periodically high turbidity (in excess of 80 NTU). In well operated prefilters, suspended solids, turbidity, and microbiological contamination can be significantly reduced. Prefilters require a "running-in" or ripening period, which may be of several months' duration for raw waters with low nutrient levels, before they reach peak operating efficiency. Care should be taken to cover the chambers or to keep water levels below the top of the gravel fill; this not only prevents birds and other animals from being attracted to the installation and fouling the prefilter, but also prevents algal growth.

In vertical upflow or downflow prefilters, periodic cleaning can be carried out by means of a high-capacity drain assembly that can be opened to allow a full filter to discharge rapidly to a waste channel. Horizontal-flow prefilters may also be cleaned in the same manner but this is less effective and the filters must periodically be emptied of gravel for cleaning; such prefilters are less cost-effective.

Prefilters will produce significant improvements in water quality when correctly designed and operated. They are particularly useful for small surface supplies when slow sand filters are overloaded with silt, and they can be managed by community caretakers if adequate support is provided by the water-supply agency. During a sanitary inspection of a prefilter the following are the principal points that should be checked:

- Is the turbidity of the water leaving the prefilter less than 60 NTU?
- Is the flow rate of water through the filter medium controlled and appropriate for local conditions (e.g. in the range 0.5–1 $m^3/m^2$ per hour)?
- Is the effectiveness of turbidity removal by the prefilter in the range 70–90% when turbidity is greater than 100 NTU?
- Is the prefilter routinely cleaned?
- Is the cleaning effective? (This may be checked by taking a sample of gravel and estimating the amount of silt present by sieve analysis.)
- Are the filter and filtrate protected from recontamination by animals and birds?

A vertical upflow gravel prefilter is shown in Fig. 6.10.

### 6.6.5 Slow sand filtration

Slow sand filtration improves the physical, chemical, and microbiological quality of water; it is reliable and inexpensive, and is therefore particularly useful in small-community water supplies.

Slow sand filters consist of a bed of sand overlying a gravel support layer and an underdrainage system. The depth of the sand filter bed is typically in the range 0.5–1.2 m, varying as the sand is skimmed off from time to time to prevent blocking on the upper surface. A sand bed depth of 0.5 m should be considered the absolute minimum to ensure adequate treatment. When a bed has been skimmed down to this depth it should be reconstructed using clean sand. The sand skimmed from the top of the bed is generally used again after it has been washed. The sand filter bed is submerged beneath supernatant (influent water) to a depth of approximately 0.6–1.5 m. Where possible, slow sand filters should be covered for protection from sunlight, which can promote the growth of algae. Covers can also reduce the risk of fouling by birds and animals and (in cold climates) of freezing.

Slow sand filters are generally operated with filtration rates in the range 0.1–0.3 $m^3/m^2$ per hour and require a much larger area than a rapid gravity filter of similar capacity. Filter sand should have a medium to coarse grading; sands containing appreciable amounts of fine particles will be quickly blocked by suspended solids in the influent flow. It is generally necessary to wash sand before using it in a slow sand filter.

The most significant feature of slow sand filtration is that the purification of the influent is effected by microbiological means. A thin, slimy mat, known as the *schmutzdecke* or filter skin, forms on the upper surface of the filter bed; this is largely organic in character and biologically extremely active. Microorganisms in the influent water are trapped and digested in the *schmutzdecke*, and are thus significantly reduced in number. Water percolating downwards passes through a biologically active zone of depth approximately 0.3–0.4 m. Fine particles are trapped on the sand grains, where microorganisms consume organic material,

**Fig. 6.10 Vertical upflow gravel prefilter**

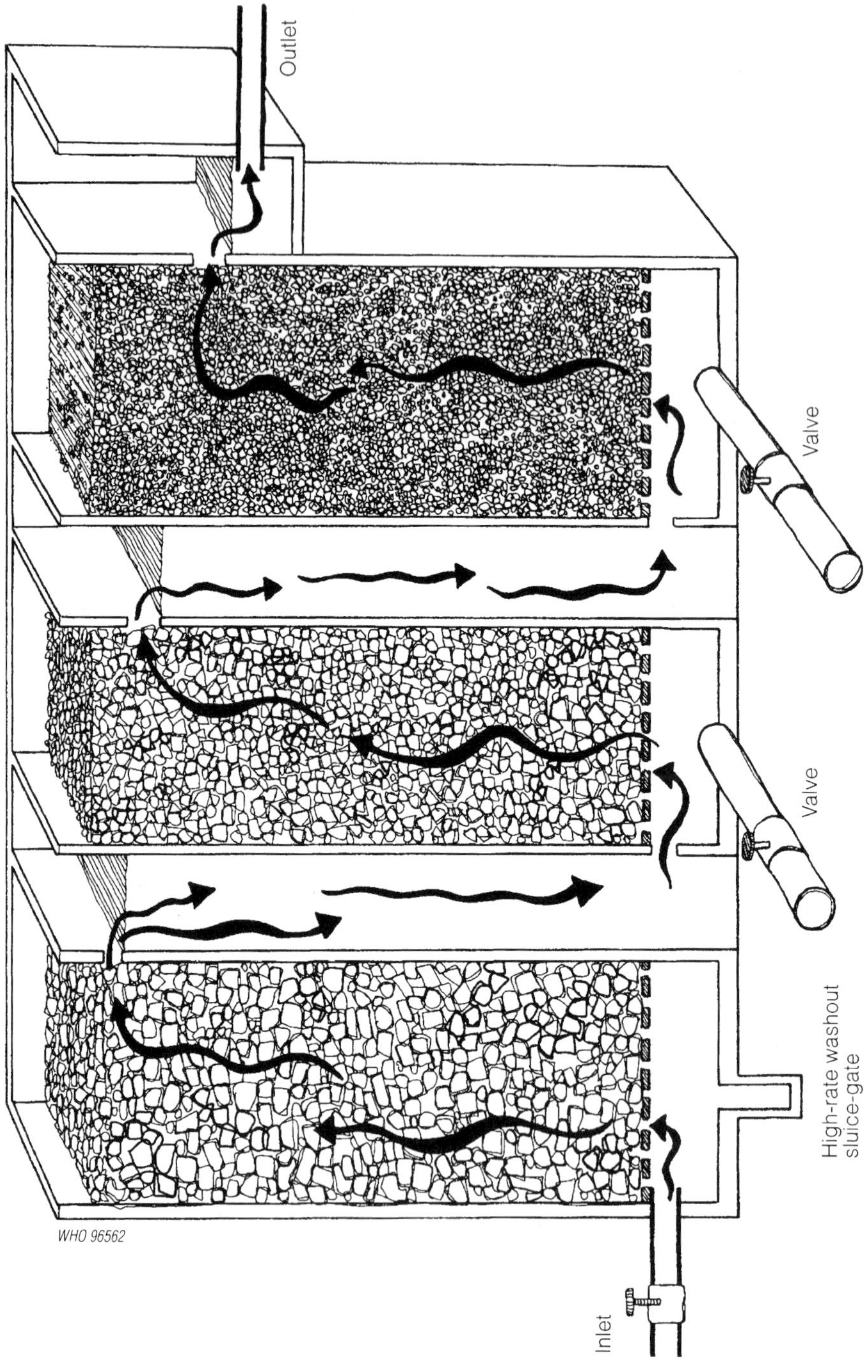

including pathogens in the influent and one another (predation). The overall effect is a substantial reduction in the number of indicator bacteria and pathogenic microorganisms in the water. In a well operated filter, the efficiency of pathogen removal may exceed 99%. The efficiency of slow sand filtration may be appreciably reduced at water temperatures below 6 °C.

After a slow sand filter is cleaned, it takes some time before the *schmutzdecke* is reestablished; with high-nutrient influent it may be a few days, but this may extend to a few weeks if the nutrient content is low. During this time, water should be allowed to flow through the filter, but it should not—ideally—be supplied to consumers. Where possible, two slow sand filters should be constructed, so that one can continue to operate while the other is being cleaned.

Slow sand filters should be operated at a constant flow rate and must never be allowed to dry out during a filtration run. Raw-water turbidity should not exceed 60 NTU for more than a few hours, since this leads to rapid blockage and consequent inefficiency in operation. Thus the efficient functioning of slow sand filters often depends on the filters being protected from high raw-water turbidities, e.g. by means of prefilters.

A typical slow sand filter design is shown in Fig. 6.11. The level of the water outlet from the filter is higher than that of the sand bed in order to avoid accidental drying of the bed due, for example, to an interruption in the source flow. Drying of the bed will rapidly kill the organisms responsible for purification.

Sanitary inspection of slow sand filters should check the following principal points:

- Is the turbidity of the filtered water less than 5 NTU?
- Is the flow rate of the water through the sand filter in the range 0.1–0.3 $m^3/m^2$ per hour and is it constant?
- Is the turbidity of the water entering the slow sand filter consistently less than 60 NTU?
- Is the slow sand filter skimmed when necessary?
- Is the depth of the sand in the filter bed greater than 0.5 m?
- Is the skimmed sand washed and stored in a sand store?
- Is a minimum head device installed and does it prevent drying of the bed if the source flow is interrupted?

### 6.6.6 Coagulation, flocculation, and sedimentation

Fine suspended particles may be removed from water by dosing with chemicals that cause formation of an absorbent, bulky precipitate. These chemicals are known as coagulants and react with suspended particles to produce settleable flocs. Most coagulants are salts of iron or aluminium, e.g. aluminium sulfate (alum) and ferric chloride. The nature of the floc depends mainly on the characteristics of the raw water, the type of coagulant employed, and the dosing rate. Rapid mixing is essential as soon as the coagulants are added to the water. After

***Fig. 6.11 Slow sand filter***

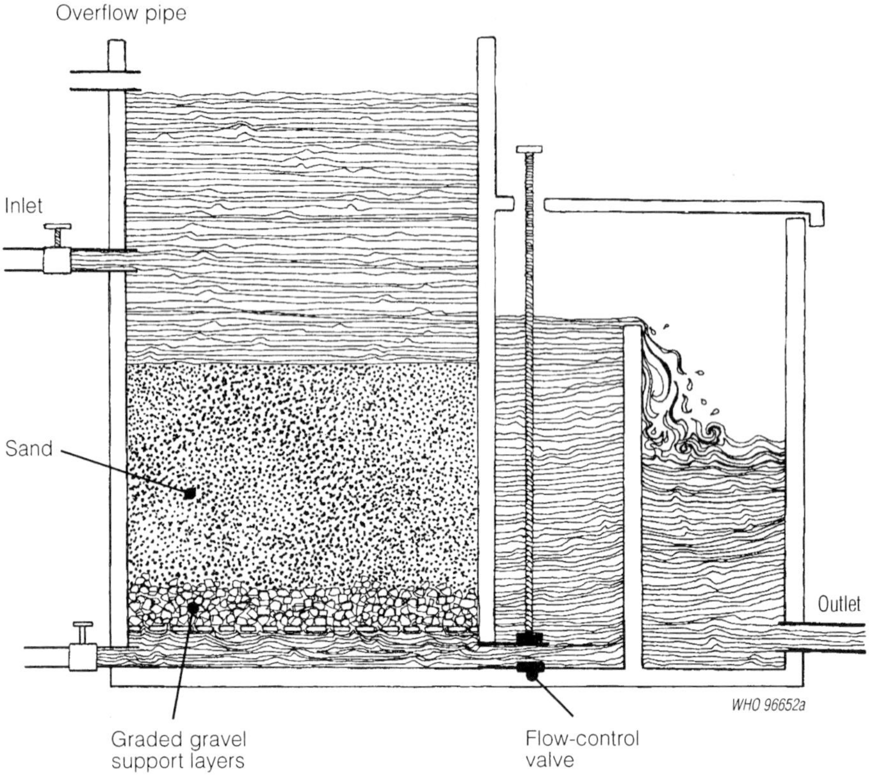


mixing, microflocs start to form which, following a suitable period in a flocculator, aggregate into settleable and filterable macroflocs. These are removed by secondary sedimentation in a clarifier, by filtration or by a combination of both processes in series. The heavier the precipitate or floc, the quicker will be its rate of settlement.

Coagulants are generally added downstream of any pretreatment such as screening or prefiltration that is designed to remove larger particles from the source water. This then allows the coagulant to act more efficiently on the finer particles.

The coagulation, mixing, and flocculation tank generally takes the form of a rectangular basin, the water flowing horizontally from one end of the tank to the other. Floc settles in the lower levels of the tank, and a high-level outlet or weir takes off the clear water (Fig. 6.12). Removal of the floc from the lower levels of the tank may be effected by means of drains. Some clarifiers are constructed in the form of an inverted pyramid, the water entering at the base and flowing upwards

*Fig. 6.12 Coagulation, mixing, and flocculation tank*

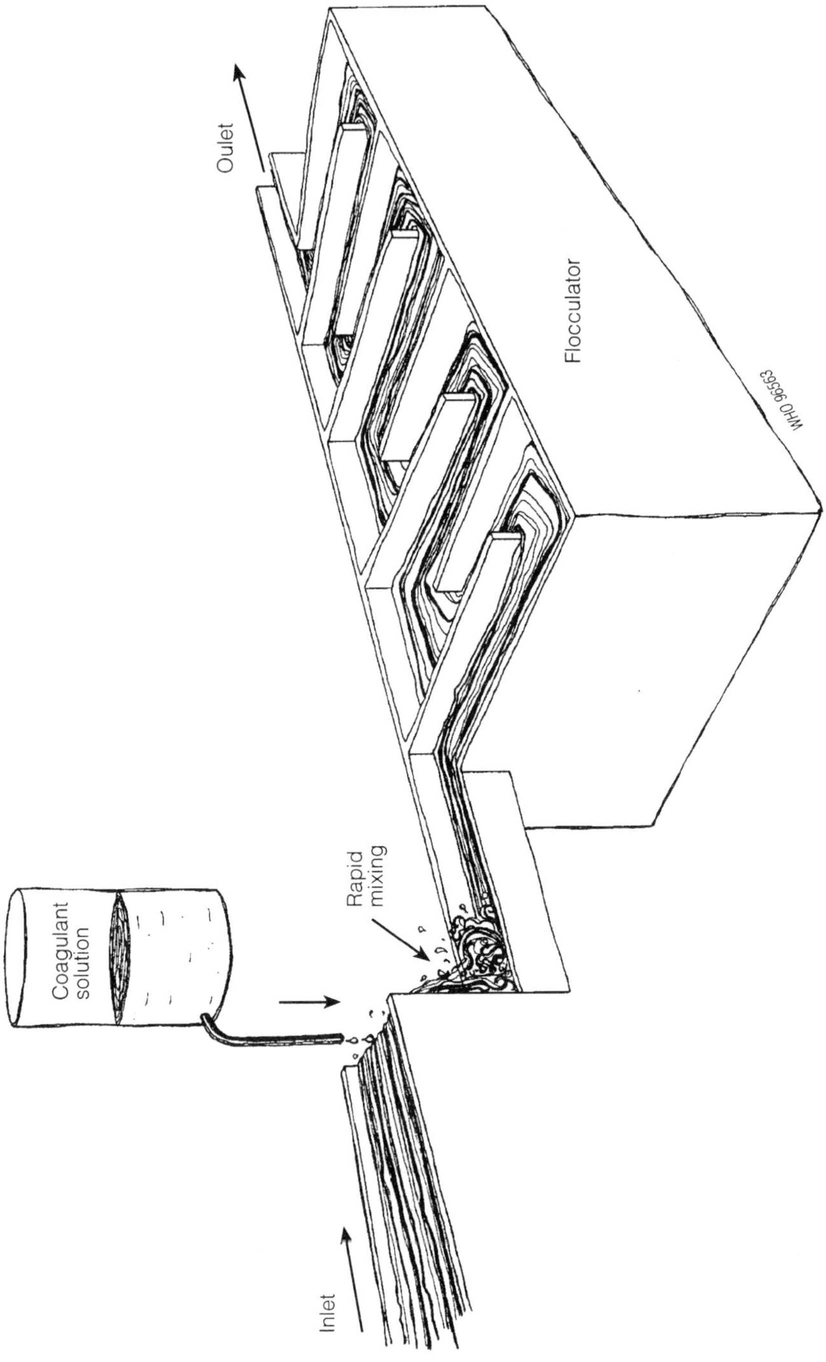

through the ever-widening tank with steadily decreasing velocity. A "sludge blanket" forms at a position where the upward force of the flow balances the downward force exerted on the floc by gravity. Clear water continues upwards, to be taken off by high-level outlets; the accumulating sludge must be "bled off" continuously to maintain the sludge blanket.

The physicochemical characteristics of the raw water determine the choice and quantity of coagulant required. These characteristics may vary with the season so that periodic adjustment of coagulant dose may be required. The problem most commonly encountered in coagulant treatment is incorrect choice of dosing rate. It is therefore essential to carry out regular jar tests to determine the optimum dose, taking into account fluctuations in turbidity or suspended solids loadings, and any other relevant factors. Quality-control procedures should also include the routine monitoring of turbidity and pH. The type and dose of coagulant can usually be determined only by experimentation in the laboratory.

During sanitary inspection, stocks of chemicals should be checked to ensure that they are safely and correctly stored, properly dispensed, used in rotation, and recorded in an inventory.

Coagulation and flocculation require relatively large financial outlay on plant, tanks, chemical dosing, and maintenance. Inevitably, therefore, the cost of any water treated in this way is high. The technique may be of some value to certain small communities, such as periurban fringe settlements, which can be easily reached by maintenance personnel from the water supplier. Coagulation may also be useful in helping to remove some chemical contaminants such as fluoride. Generally, however, the technique is too difficult to apply and control satisfactorily in most isolated rural communities.

A sanitary inspection check list is included in Annex 2.

### 6.6.7 Rapid sand filtration

In large treatment works, rapid sand filtration is frequently used after coagulation–flocculation–sedimentation and before disinfection. It may also be used as a prefiltration step before large-scale slow sand filtration. Rapid filtration can be carried out in open tanks (rapid gravity sand filters) or closed metal tanks through which the water passes under pressure (pressure filters). Rapid gravity filters usually operate at filtration rates considerably higher than those typical of slow sand filtration (about 4.0–5.0 $m^3/m^2$ filter area per hour). As a consequence, the filters are considerably smaller in area for a similar throughput capacity. Coarse sand is generally used in rapid gravity filters; multimedia filters (containing e.g. very coarse anthracite particles above coarse sand) have been employed where it has been necessary to protect against blocking of the surface of the filter by straining. Rapid gravity filter beds are generally 0.6–1.0 m in depth with typical particle diameters in the range 0.4–1.0 mm.

Microbial removal rates in rapid gravity filters are low, but suspended solids are removed quite efficiently. Filters are quickly blocked by surface straining or excessive sedimentation in their upper layers. Cleaning must therefore be carried out regularly (typically daily), and involves vigorous backwashing with water, sometimes in combination with compressed air scour. When rapid gravity filters are overloaded, breakthrough can occur within a very short time because of the coarse nature of the media employed. If overloading is a problem, an increase in backwashing frequency or plant capacity will be required. Mudballing and cracking can occur in the filter bed if routine cleaning is not carried out in a proper and effective manner.

### 6.6.8 Aeration

Aeration can be used in water treatment to reduce tastes and odours (e.g. by oxidation of hydrogen sulfide), lower the levels of volatile organics, and alter the concentrations of dissolved gases, although it has little appreciable effect on those associated with algal growth. The aerators best suited for use in community supplies are the cascade, multiple-tray, and packed-bed types, in which a thin film of water flows over surfaces to maximize oxygen transfer into the water from the surrounding air.

#### *Cascade aerator*

A cascade aerator consists of a stairway over which water flows in a very thin film. Typically, the width and depth of each step is 10–15 cm and the height 1–4 m; the head requirement for larger cascades can be a major design problem if pumping is to be avoided.

#### *Multiple-tray aerator*

A multiple-tray aerator (see Fig. 6.13) comprises a series of trays formed by perforated metal plates, metal screens, or wooden slats, arranged vertically above one another in the form of a small tower. The individual trays contain a layer 15–30 cm deep of stone, coke, or fired-clay material 5–15 cm in size. Water is delivered to the top of the multiple-tray assembly where it is either sprayed or sprinkled from a perforated tank onto the top tray. Appropriate loading rates should be determined by pilot plant trials, as the characteristics of water can vary from one site to another and may also be subject to seasonal changes. Typical loading rates are in the range 0.25–10 $m^3/m^2$ of total effective tray area per hour.

Air passes through the media, and open louvres are often inserted between the trays to maximize the flow. In some extreme circumstances, mechanical forced-draught ventilation may be employed to maintain the highest possible rate of aeration in a particular installation. Performance can be badly affected by ice formation during periods of freezing weather.

Apart from aeration, multiple-tray aerators can also be used to remove the iron present in some waters.

***Fig. 6.13 Multiple-tray aerator***

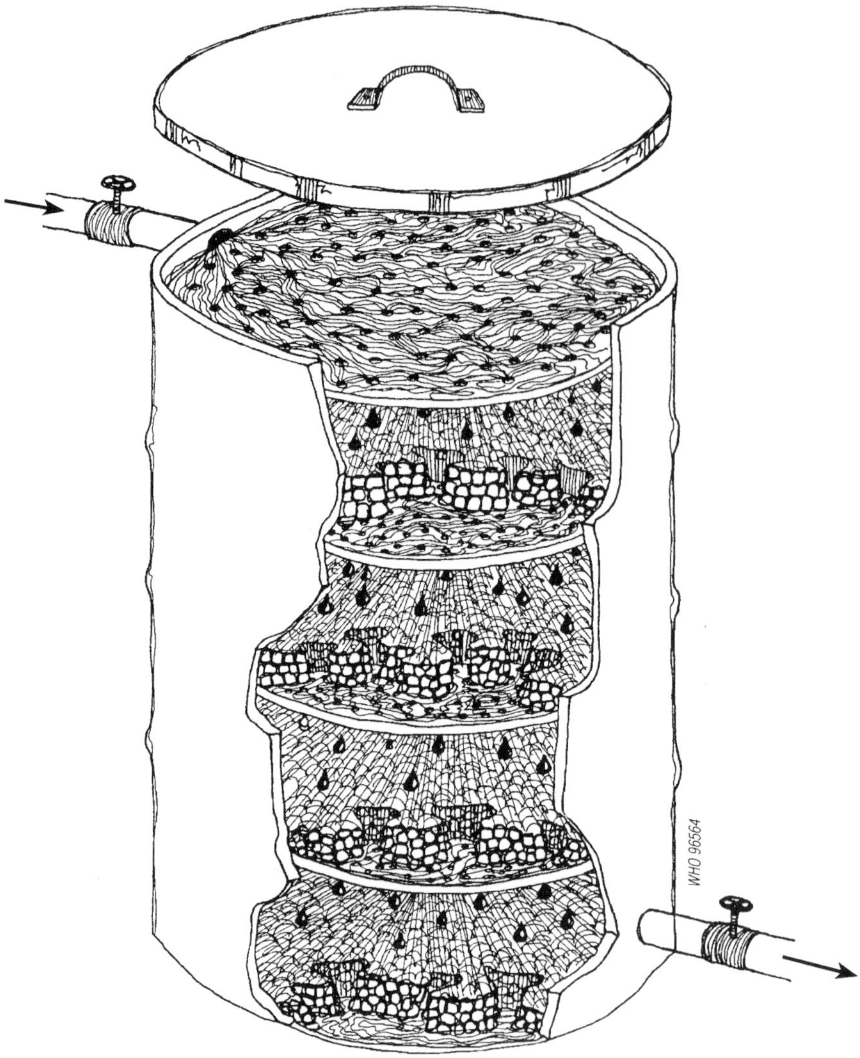


### *Packed-bed aerators*

A packed-bed aerator consists of a tower containing fired-clay, ceramic, plastic, stone, or coke media of particle size 5–15 cm and is generally used to strip volatile organics from the water stream. Specialized media are available, including ceramic cylinders or plastic in various shapes. Forced ventilation is required, and the performance must be determined in a pilot plant before a full-scale installation is constructed.

If the source water is rich in metals (some groundwaters contain iron, for example), concentrations above 0.3 mg/litre may produce detectable taste and odour. Furthermore, water containing iron may cause stains when used for laundry, accumulations of iron precipitates in the pipework of the distribution system, and the growth of *Crenothrix* bacteria. Packed-bed aerators can be used to remove iron, which is deposited on the media. Manganese removal is more difficult to accomplish and must be carried out at a pH greater than 9, and combinations of metals can also be difficult to remove. The addition of strong oxidizing agents, such as chlorine, ozone, or potassium permanganate, can assist in the deposition process.

As with multiple-tray aerators, the performance of packed-bed aerators can be badly affected by ice formation if periods of freezing weather are experienced.

### 6.6.9 Fluoride removal

Fluoride can occur naturally or may be added to drinking-water during treatment. A fluoride concentration of around 1 mg/litre can help to reduce the incidence of tooth decay, but concentrations above 1.5 mg/litre may cause browning of teeth; very high concentrations may cause skeletal fluorosis.

High fluoride levels, for example in groundwaters, are locally common in some areas of the world, and in most such circumstances it may be more practical and cost-effective to use alternative water sources. However, fluoride can be removed from water by filtering through bone char, which can subsequently be regenerated, and this approach has been adopted for some small-community water supplies.

Addition of fluoride to drinking-water supplies to reduce the incidence of dental caries should be closely monitored to ensure that safe levels are not exceeded. The fluoride is generally added in the form of a solution, both for convenience and because powders are toxic and require special handling arrangements. Hydrofluosilicic acid provides a suitable solution for this purpose, although the normal precautions required in the handling of acids must then be taken and appropriate equipment is required.

### 6.6.10 Control of nitrites and nitrates

The presence of either nitrites or nitrates in drinking-water is a matter of concern from the point of view of human health, since there is evidence that they may cause methaemoglobinaemia in infants. Nitrites and nitrates are present in surface waters mainly as a result of the oxidation of ammonia in sewage effluents and the excessive use of nitrate fertilizers in farming. Nitrite can occur as an intermediate stage in the oxidation of nitrogen to nitrate. Nitrates in groundwaters are often reduced to nitrites.

Algal assimilation can significantly reduce nitrate levels in surface waters. Seasonal variations in nitrate levels in rivers and streams are likely to occur for reasons associated with changes in the overall levels of biological activity in the water.

There is no water treatment method for reducing nitrite and nitrate levels that is both convenient and generally appropriate for small-community water supplies. Consideration should therefore be given to the protection of water sources, particularly where the principal sources of contamination are the agricultural use of fertilizers or wastewater and sewage discharges. If seasonally high levels are experienced in a river source, it may be possible to blend water from lake or groundwater sources with the surface water to achieve the required quality. Bankside reservoir storage can provide an opportunity to close intakes when high peaks in river nitrate levels are expected. Algal activity in reservoirs can reduce nitrate levels significantly, aided by the denitrifying activities of bacteria in the bottom silt layer.

### 6.6.11 Disinfection

The microbiological quality of drinking-water can be substantially enhanced by protecting the source and by treating the raw water, especially if slow sand filtration is employed. However, where raw waters are not of a consistently high quality, some form of disinfection is essential to ensure that the supply is microbiologically safe. Provided that the physical and chemical quality of the water is acceptable, disinfection provides the most effective means of reducing the numbers of microorganisms in drinking-water.

Disinfection methods may be either physical or chemical. Physical methods include boiling and ultraviolet (UV) irradiation; chemical methods include the addition of ozone, or, most commonly, chlorine and its derivatives. Only chlorination has been widely applied in treating community water supplies, although UV irradiation is also sometimes appropriate, as is on-site generation of disinfectant gases.

Chlorine is an oxidizing agent that reacts rapidly with organic and inorganic matter present in water. If adequate disinfection is to be achieved, due allowance must be made for the chlorine consumed in these reactions in addition to that needed for disinfection. The amount of chlorine required to react with other compounds (mainly ammonia, some metal ions, and organic compounds) is termed the chlorine demand of the water. Thus, the chlorine dose must be sufficient both to satisfy the chlorine demand and to produce an unreacted excess known as the free residual. A minimum free residual of 0.5 mg/litre is recommended, together with a minimum contact time of 30 minutes and a water turbidity of less than 5 NTU (ideally less than 1 NTU). The chlorine demand of some waters (particularly river waters) can increase dramatically at times of heavy pollution, particularly after rain. It may therefore be necessary to increase the dose to allow for this. The residual chlorine level should be determined (see Annex 9) in samples taken from various points throughout the distribution system, to ensure that a free residual exists in the water supplied to the public.

Chlorination usually requires the addition of one of the following three substances to the water:

- Chlorine gas, $Cl_2$, liquefied under a pressure of 505kPa (5atm). This requires careful handling because it is highly toxic: the gas supplier should provide clear operational guidelines and the surveillance officer should check that these are being strictly observed.
- Sodium hypochlorite solution for water disinfection, containing up to 14% available chlorine, or liquid bleach (about 1% available chlorine). Solutions are unstable at warm temperatures and should be stored in brown or green glass bottles or opaque plastic bottles in a cool, dark place. They should be checked regularly to ensure that the chlorine content is adequate since the concentration may fall if the container has been opened or stored for a long time.
- Solid calcium hypochlorite, commonly available as bleaching powder or chlorinated lime, containing about 30% available chlorine when fresh. The compound is unstable at warm temperatures and should be carefully stored. High-test hypochlorite (HTH) can also be used; it normally contains 50–70% available chlorine.

Simple devices for use in chlorination include the constant-head drip and double-pot chlorinators; typical examples are shown in Figs 6.14 and 6.15, respectively.

### 6.6.12 Water-treatment plants

The only proven method of treating polluted surface water by means of simple equipment is based on the multiple-barrier principle, i.e. on the use of at least three unit treatment processes in series which progressively remove pathogens and other contaminants (notably turbidity). The technology is robust and has the advantage that failure of any one barrier should not significantly increase the risk of transmission of infectious waterborne disease. A typical multiple-barrier series of unit processes is shown schematically in Fig. 6.16, and includes:

— plain sedimentation
— triple-stage gravel prefiltration
— slow sand filtration
— disinfection.

The two main parameters determining the selection and performance of treatment plants are the thermotolerant (faecal) coliform count per 100ml and the turbidity. These should be reduced so that, however many unit processes are employed, the water leaving the plant always has a zero thermotolerant (faecal) coliform count and turbidity below 5 NTU. These treatment objectives have been incorporated into Table 6.2 to show the required performance of the unit processes considered appropriate for community water supply.

***Fig. 6.14 Constant-head drip chlorinator***

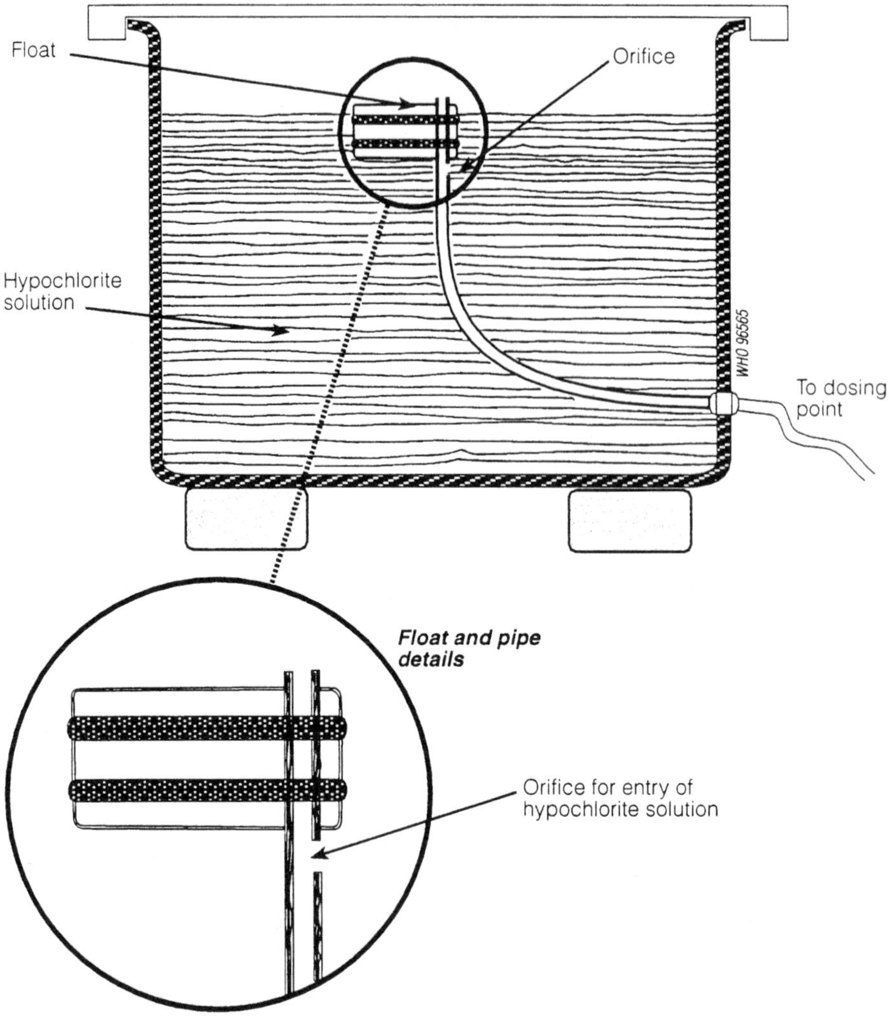


## 6.7 Household water treatment and storage

Where the source of water used by a community is unprotected and/or untreated, or when the water supply is contaminated, household water may require treatment in the home to ensure that it is safe for consumption. Household treatment and hygienic storage can improve the aesthetic quality of water (turbidity, temperature, etc.) and reduce faecal contamination, but its use to improve chemical

***Fig. 6.15 Double-pot chlorinator***

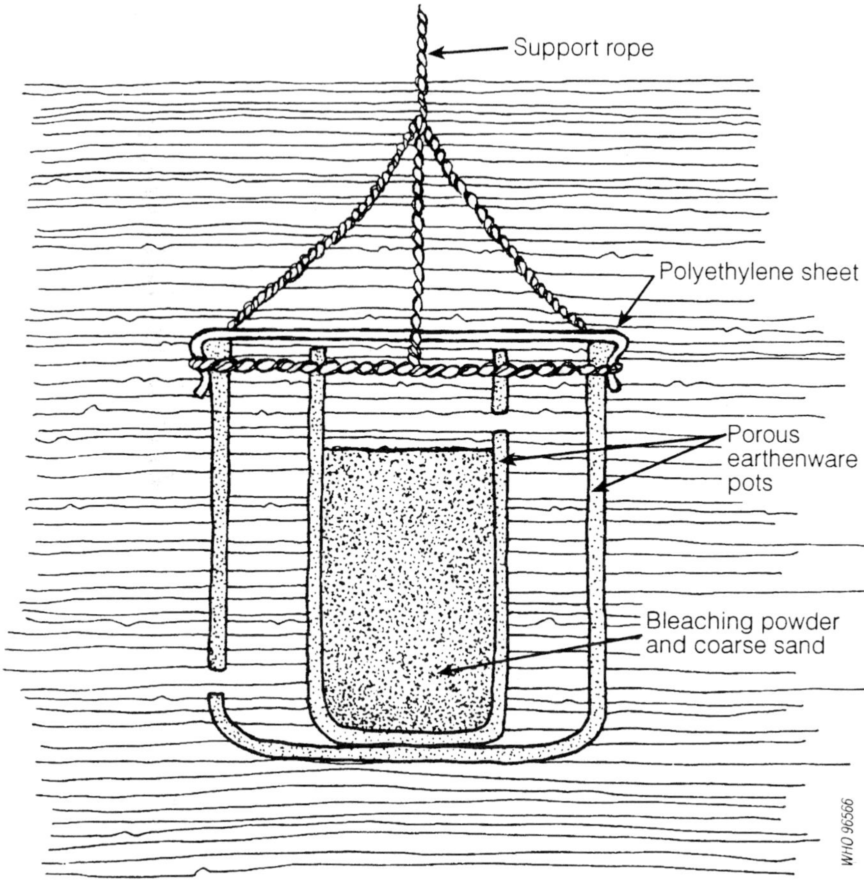


***Table 6.2 An example of performance objectives for removal of turbidity and thermotolerant coliform bacteria in small-scale water treatment***

| **Stage and process** | **Turbidity** | | | **Thermotolerant coliform bacteria** | | |
|---|---|---|---|---|---|---|
| | **Removal (%)** | **Average loading (NTU)** | **Maximum loading (NTU)** | **Removal (%)** | **Average loading (per 100 ml)** | **Maximum loading (per 100 ml)** |
| Plain sedimentation | 50 | 60 | 600 | 50 | 1000 | 10 000 |
| Gravel prefilters (3-stage) | 80 | 30 | 300 | 90 | 600 | 5000 |
| Slow sand filter | >90 | 6 | 60 | 95 | 50 | 500 |
| Disinfection | NA[a] | <1 | <5 | >99.9 | <3 | 25 |
| Distributed water | NA[a] | <1 | <5 | NA[a] | <1 | <1 |

[a] NA, not applicable.

**Fig. 6.16 Typical multistage treatment system for small-community supplies**

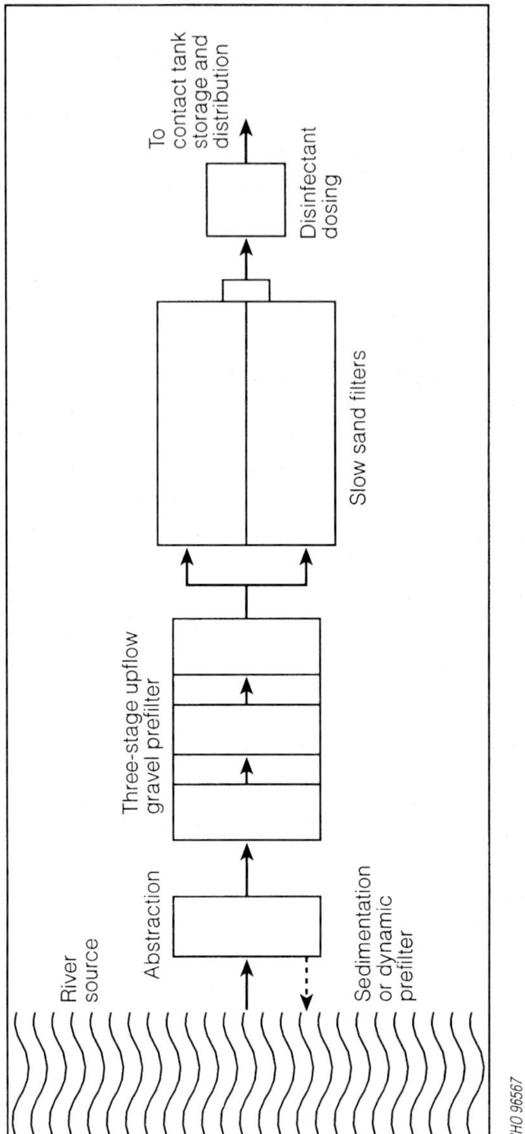

quality is uncommon; this section therefore deals only with reducing the faecal contamination of drinking-water to prevent the transmission of infectious waterborne diseases.

In many situations water must be transported, often carried, from a well, spring, or standpost to households. In these circumstances and where the water supply to the household is intermittent, water must be stored in the home to ensure that enough is available when it is needed. Water that is transported or stored unhygienically may be recontaminated, which represents a public health risk; water supplied at the well or standpost may be microbiologically safe but become grossly contaminated with faecal material before consumption because of poor handling. A surveillance programme should therefore include the testing of water stored in the household to establish whether recontamination is occurring.

If drinking-water regularly becomes recontaminated, the best remedial action is a hygiene education programme. This should involve all the community but focus particularly on those members with most responsibility for water collection, storage, and treatment (usually women and children). Most recontamination is the result of behavioural patterns; if these can be changed, the health risk can be reduced or eliminated. Technical interventions (like those described below) may also be used, but are unlikely to result in any significant reduction in recontamination without a complementary hygiene education programme. Hygiene education is dealt with in Chapter 7.

### 6.7.1 Household water treatment

Where local water supplies are known to be contaminated or have not been tested, household treatment should generally be recommended. Faecally contaminated water can be treated by:

- — boiling
- — filtration
- — chemical disinfection
- — cloth filtration (to prevent dracunculiasis).

#### *Boiling*

Boiling is a simple way of killing any ova, cysts, bacteria, and viruses present in contaminated water. Water should be heated until it comes to a "rolling boil" (large bubbles continuously coming to the surface of the water) which is maintained for 1 minute. Water boils at a lower temperature as altitude increases, and 1 minute of extra boiling time should therefore be added for every 1000 metres above sea level. Boiling has the following disadvantages:

- Large amounts of fuel are required, so that cost may prevent people from boiling water in many areas.
- It may give an unpleasant taste to the water which may be unacceptable.

- Very hot water can cause accidents in the home.
- Boiled water can become recontaminated once it has cooled.

### Simple household filters

There are many different types of household filter, some produced commercially and others that can be manufactured locally. Most will remove a high proportion of solids and silt. Many will also remove parasites including cysts, ova, and guinea worm larvae, but some simple filters may not remove all microorganisms from water. The various types of simple household filter are candle, stone, and sand filters.

*Candle filters* are often commercially produced. In this type of filter, contaminated water is allowed to filter slowly through a porous ceramic material (see Fig. 6.17). Larger microorganisms—ova, cysts, and most bacteria—are left in the outer layer of the filter material, which is periodically cleaned by gently scrubbing the filter under clean, running water. Smaller microorganisms, such as the virus that causes hepatitis A, may not be removed by candle filters.

Candle filters should be designed to minimize the risk of recontamination of water after filtering. Most commercial filters consist of two interlocking containers. The upper container for the candle(s), into which the raw water is poured, is usually fitted with a lid. The base of this container fits securely onto the top of the lower container; an overlapping lid prevents recontamination of the filtered water. The lower container, which collects the filtered water, is fitted with a tap near the base to allow hygienic withdrawal of the water.

It is important that the manufacturer's instructions for cleaning and the safe life span of the filter should be carefully followed.

*Stone filters* are similar to candle filters but are carved from porous local stone (see Fig. 6.18). They are generally difficult to clean and heavy to lift, but have the advantage of being relatively inexpensive if they can be produced locally. If these filters are commonly used in a particular area, it would be worthwhile to test water from a representative sample to determine the efficiency of removal of faecal contamination. Filtered water is generally collected in an open vessel, often close to the ground, so that there is a significant risk of recontamination.

*Sand filters* should not be confused with the slow sand filters discussed earlier in this chapter, which are very efficient at removing microorganisms from contaminated water. A slow sand filter would be difficult to operate in a household as it requires a continuous and constant flow of water if it is to function effectively. Household sand filters (see Fig. 6.19) will remove solid material from water and often ova, larvae, cysts, and *Cyclops* spp. Because bacteria and viruses are not removed, additional treatment, such as disinfection (usually with chlorine), may be desirable after filtration.

*Removal of turbidity.* When water is extremely turbid, it may be necessary to remove some of the particulate matter before the water is passed through a filter in order to avoid blockage. Pretreatment, either by settling or coagulation, will often also help to reduce faecal contamination to some extent.

- Settling. If turbid water is left in a closed container for several hours, e.g. overnight, a proportion of the suspended matter will settle to the bottom. The clearer water can then be decanted from the top and poured into a filter.
- Coagulation. Turbid water can be settled more rapidly and effectively if a chemical coagulant is used to make the suspended particles stick together.

***Fig. 6.17 Candle filter***

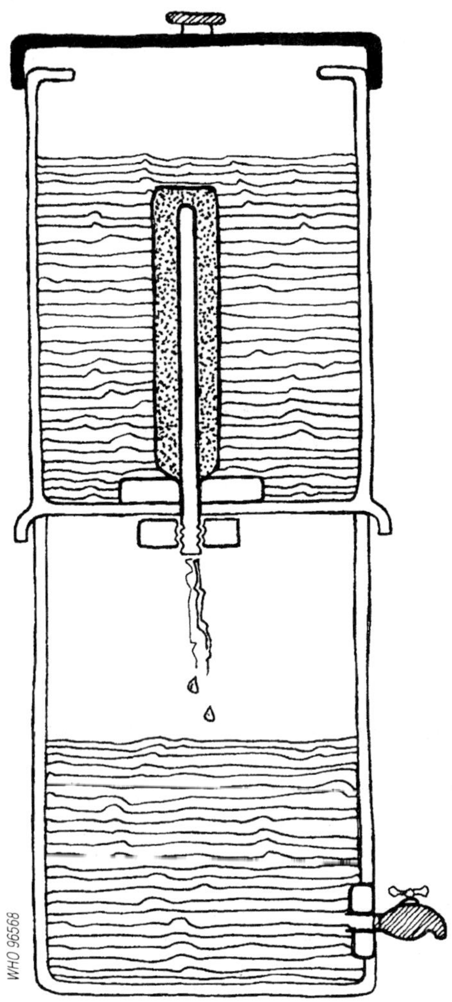

The dose of alum required will depend on the turbidity of the water and should be selected on the basis of local experience whenever possible. Certain indigenous plants can also be used to make suspended particles stick together, and in some areas such natural coagulants are widely and successfully used. So many different plants are used for this purpose in different parts of the world that no general recommendations can be made. Local experience and practice should be investigated and used as a guide.

***Fig. 6.18 Stone filter***

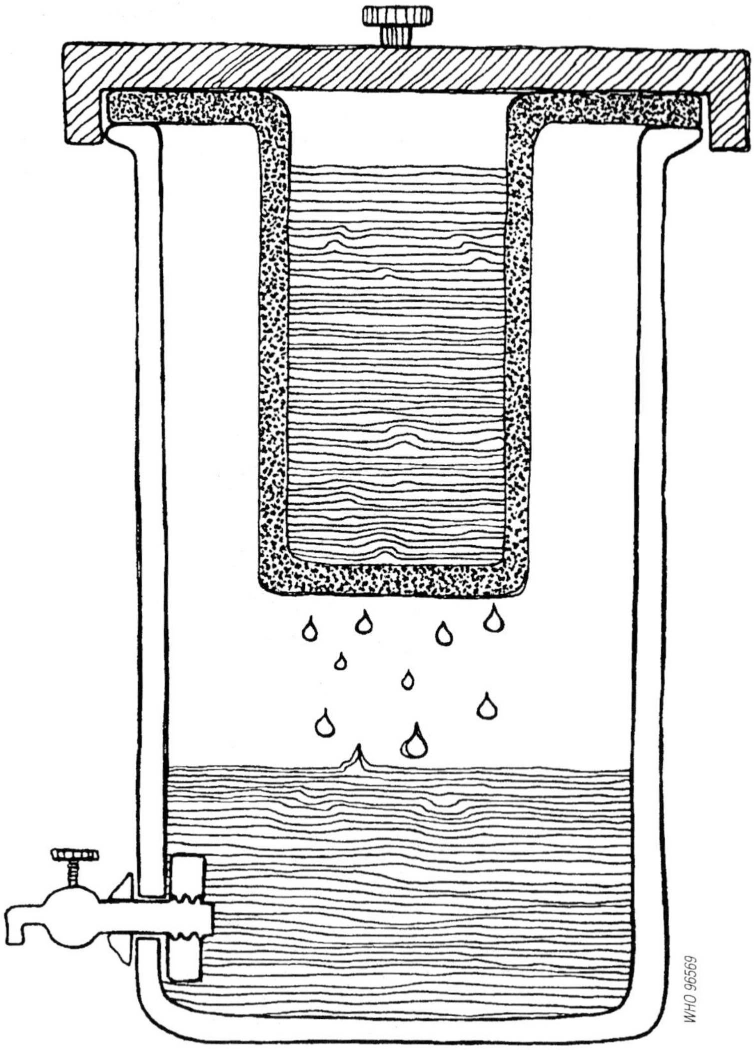

### Disinfection

If water is contaminated but clear, disinfection can be used to kill the microorganisms it contains. Using chlorine for this purpose will provide a disinfecting residual that will help to prevent recontamination.

Of the various ways of disinfecting household drinking-water, the commonest is to use chlorine. A 1% solution of chlorine is often used, in the form of sodium hypochlorite (liquid bleach), calcium hypochlorite (generally as a powder), or HTH (high-test hypochlorite in powdered form); see also p. 115.

*Chlorine is a hazardous substance. It is highly corrosive in concentrated solution and splashes can cause burns and damage the eyes.* Appropriate precautions should be taken when concentrated chlorine solutions or powders are handled. If the eyes or skin are splashed, they should immediately be rinsed thoroughly with water. Solid forms are less hazardous to handle during transport than solutions. It is good practice to wash the hands after handling concentrated chlorine in any form. All containers in which chlorine is stored should carry a label clearly identifying the contents and including a hazard warning in a form that is readily understandable locally. Storage sites for chlorine in any form should be secure, and special precautions should be taken to prevent access by children.

**Fig. 6.19 Household sand filter**

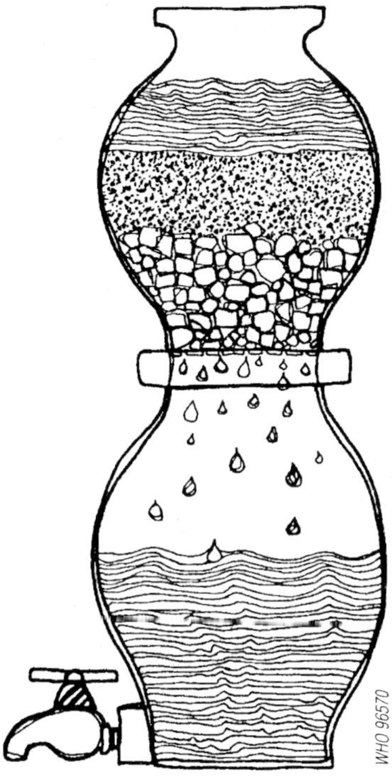

Where drinking-water is to be both disinfected and filtered, disinfection should follow filtration; otherwise the disinfectant may be neutralized by the filter. Disinfection is less effective in turbid or cloudy water as the chlorine can be consumed by the suspended particles in the water; particulate matter may also protect bacteria from the disinfectant action of chlorine.

***Fig. 6.20 Method of preparing chlorine solutions using local materials***

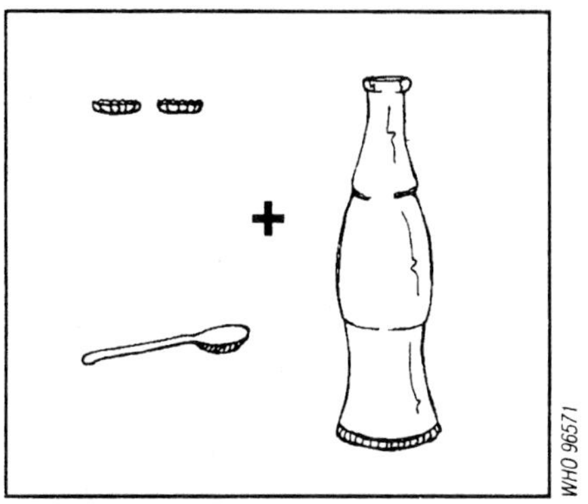


1. Fill two bottle tops, or one level teaspoon with chlorine powder (HTH), put into a small drink bottle (about 300 ml) and add clean water to the top.

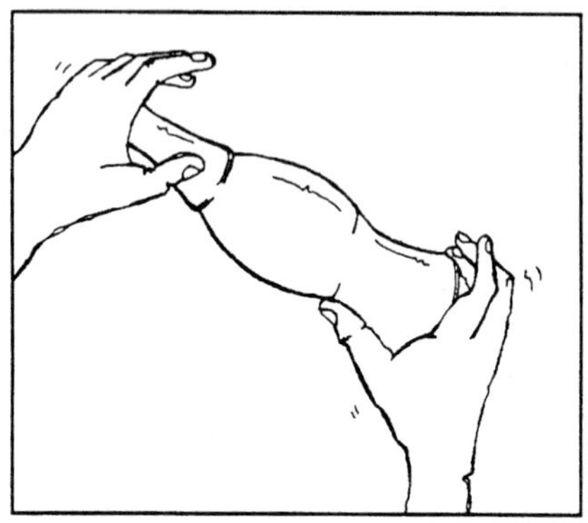


2. Cork the bottle and mix well for 2 minutes. Leave to stand for 1 hour.

3. Now you have the same chlorine as household bleach. Put it in a dark place away from children.

4. Add 3 drops of the chlorine solution for every litre of water. Leave for 1 hour, then taste. You should just be able to taste the chlorine. If you cannot taste it, add 1 drop per litre until you can. The water will only be safe to drink for 24 hours.

Sodium hypochlorite solution can be used directly to disinfect household drinking-water as its chlorine concentration is already 1%; calcium hypochlorite and HTH need to be diluted to this concentration before use. The quantity of powder used will depend on the concentration of chlorine present. Community members should employ locally available and familiar containers and units of measurement. An example of a method of preparing chlorine solutions which has been used successfully is shown in Fig. 6.20.

#### ***Cloth filtration to prevent guinea-worm disease***

Guinea-worm disease (dracunculiasis) is transmitted via contaminated drinking-water (e.g. from stagnant ponds, cisterns, or step wells). The disease occurs in a number of countries in Africa and Asia and causes severe suffering and disability among the world's most deprived people. Infected individuals do not develop immunity. There is no known animal reservoir, and people can disseminate the parasite 1 year after infection and during 1–3 weeks after emergence of the worm. For these reasons and because control of transmission, including treatment of drinking-water, is simple, global eradication of this disease is feasible.

Dramatic reductions in the prevalence of dracunculiasis have been achieved through improvement of water supplies and by promoting proper hygiene in areas where the disease is endemic. In such areas, guinea worm (*Dracunculus medinensis*) can be effectively eliminated by filtering all drinking-water through fine cloth (see Fig. 6.21). Filtration of drinking-water is thus a primary strategy for the control of guinea-worm disease.

Filters should be of mesh size less than 130 μm; this should remove all infected intermediate hosts. Monofilament synthetic cloth (nylon) is most suitable because it clogs less rapidly and is easily cleaned; it has a mesh size of 100–130 μm. Cotton cloth can be used but tends to clog rapidly. Boiling is also effective as a means of controlling the disease.

### 6.7.2 Household water storage

The principal health risk associated with household water storage is the ease of recontamination during transport and storage, particularly where the members of a family or community do not all follow good hygiene practice. Good hygienic measures include the following:

- careful storage of household water and regular cleaning of all household water-storage facilities;
- construction, proper use, and maintenance of latrines;
- regular hand-washing, especially after defecation and before eating or preparing food;
- careful storage and preparation of food.

Water that is clean from the supply or has been treated in the household needs to be protected from recontamination. The following precautions and considerations are important:

***Fig. 6.21 Pouring water through a monofilament filter to control transmission of guinea worm***

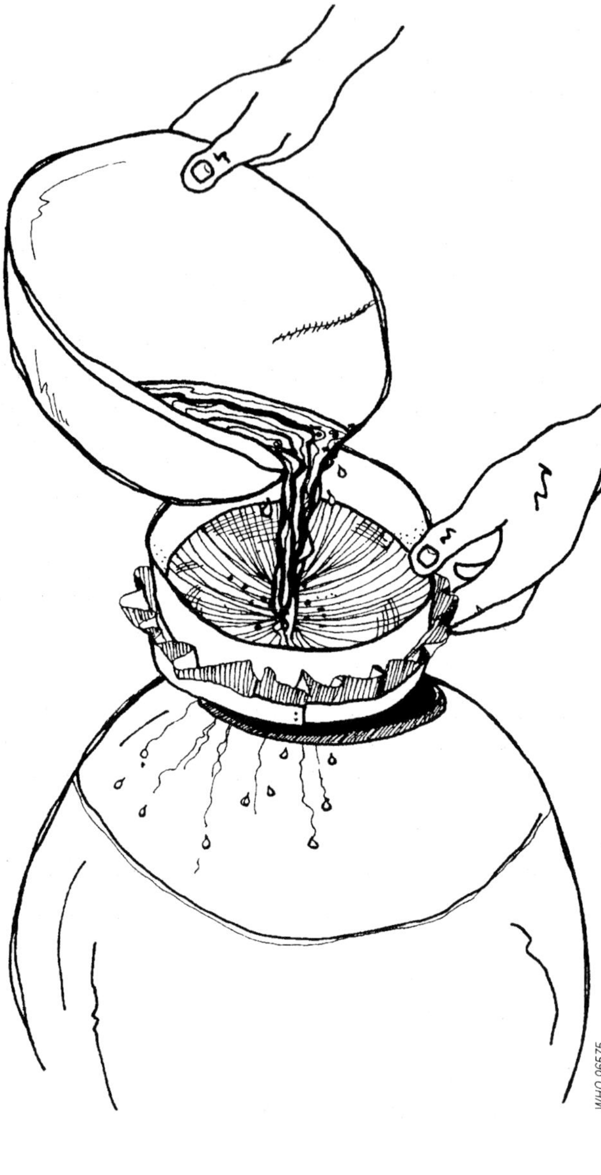


- *Location of storage vessel.* The storage vessel should be placed above ground level to restrict access by children and animals. It should preferably be placed in a shaded position to keep the water cool, and should be accessible to users and for refilling.

- *Design of storage vessel.* The storage vessel should be designed to reduce the risk of contamination: it should have a secure, tight-fitting lid, be robust enough to withstand rough handling without cracking, and be easy to lift from the ground and carry back to the storage point after filling. Stored water may be kept cool by using earthenware jars or pots; these allow some water to evaporate, which has a cooling effect. Containers should be easy to fill and clean, so that contact with hands is minimized.
- *Removal of water.* It should be possible to remove water from the container hygienically, with no contact between hands and the water. Water is commonly withdrawn by means of a cup. This may be acceptable where the cup is not used for any other purpose, is cleaned regularly, and is stored where contamination cannot occur. However, as it is difficult to dip the cup into the water without also putting in the hands, the risk of contamination is still high. It is better to use a ladle that is stored permanently inside the container; this reduces the risk of contamination while the ladle is not in use. However, the ladle should be used *only* to transfer water to a cup or other vessel. Drinking from it directly may cause contamination of the water. The ladle should be held only by the top of the handle and not by the scoop or any part that is immersed in the water during storage. Fitting a tap to the container minimizes contact with the water and is the most hygienic method of withdrawal. However, users must not wipe the tap with dirty hands or hang cups, etc. from the tap as this increases the risk of contamination before consumption. Taps are expensive, may be difficult to fit on traditional containers, and may also weaken the container.

Substances such as petrol, diesel fuel, pesticides, and solvents should not be stored or used near water facilities (sources, catchments, storage tanks, etc.). Containers that have been used for the storage, transport, or handling of these substances should not subsequently be used to store water intended for human consumption, even after thorough cleaning.

The most important elements of water storage can be summarized as follows:

- Use a clean water source or treat the water, either at home or in a storage tank.
- Store water in an earthenware or plastic container with a lid.
- Store the water container at a height that puts it beyond the reach of children and animals.
- Fit a tap to the container for drawing clean water in order to prevent contamination by dirty cups, ladles, or hands.

### 6.7.3 Storage tanks

Where a piped water supply to the household operates intermittently, a storage tank is commonly used to ensure that there is sufficient water for the family needs throughout the day. The tank should be covered to prevent contamination of the water and to restrict access by children and animals. It may be located inside or outside the house, but a secure cover should be fitted to an outdoor tank.

If the water running into the tank is clean (i.e. comes from a protected source or a treatment plant), the tank should be inspected, cleaned, and disinfected at least once a year. Where the water supplied is not clean, the tank will require more frequent cleaning, the frequency depending on the water quality. Water of poor quality should be treated by the most appropriate means.

The pipes running from a household storage tank to the taps must *not* be made of lead, which is toxic; pipes made of galvanized iron, copper, or plastic (such as potable grade PVC) should be used instead. Galvanized iron pipes should not be used where the water supplied is highly acidic or alkaline because they will corrode. A non-lead solder should be used, where possible, to join metal pipes, and a nontoxic solvent cement for plastic pipes. The system should be thoroughly flushed before use to remove any traces of solvent or metal solder from the pipes.

When a household storage tank and pipes for drinking-water are installed, they should ideally be filled with water containing 50 mg/litre of chlorine and left to stand overnight so that the system is disinfected before use.

# 7. Hygiene education

## 7.1 Scope of hygiene education

### 7.1.1 Community-based surveillance

Effective and sustainable programmes for the surveillance of water supplies require the active support of local communities, which should be involved at all stages in such programmes, including initial surveys; monitoring and surveillance of water supplies; reporting faults, carrying out maintenance, and taking remedial action; and supportive actions including sanitation and hygiene practices. This will involve setting up a comprehensive educational programme to ensure that the community:

- — is aware of the importance of water quality and its relation to health, and of the need for safe water supplies;
- — accepts the importance of surveillance and the need for a community response;
- — understands and is prepared to play its role in the surveillance process;
- — has the necessary skills to perform that role.

### 7.1.2 Hygiene behaviours

The provision of a good drinking-water supply alone is insufficient to ensure health. There are many stages in the collection, storage, and handling of food, the disposal of excreta, and the care of children at which drinking-water can become contaminated and the community exposed to pathogens in excreta.

Children, especially those under 5 years of age, are particularly vulnerable to diarrhoea. A common belief is that children's faeces are harmless, whereas in fact they are the main source of infection of other children. Parents may not hygienically dispose of their young children's faeces, young children may not use latrines, and the yards surrounding homes are often contaminated.

There are many transmission routes for water-related and sanitation-related diseases, and hygiene education can therefore cover a wide range of actions. The most important behaviours from the point of view of health will depend on the community, the disease pattern, and the climate. One of the functions of the initial field inspection and surveillance (see Chapters 1 and 2) is to determine which behaviours the hygiene educational programme should seek to promote in the community (see Table 7.1).

### *Table 7.1 Behaviours to be recommended in hygiene education*

**Water source:**
- All children, women, and men in the community should use safe water sources for drinking and food preparation.
- Adequate water should be used for hygiene purposes such as bathing, household cleanliness, and clothes washing.
- Water should be efficiently used and not wasted. Wastewater should be properly drained away.
- Improved water sources should be used hygienically and be well maintained.
- There should be no risk of contamination of water sources from nearby latrines, wastewater drainage, cattle, or agricultural chemicals.

**Water treatment:**
- Simple purification procedures, e.g. chlorination, should be carried out on the water source if necessary.
- If necessary, water should be filtered to remove any solid material, guinea worm, etc. (see section 6.7.1).

**Water collection:**
- Drinking-water should be collected in clean vessels without coming into contact with hands and other materials.
- Water should be transported in a covered container.

**Water storage:**
- Water should be stored in vessels that are covered and regularly cleaned.
- Drinking-water should be stored in a separate container from other domestic water wherever possible.

**Water drinking:**
- Drinking-water should be taken from the storage vessel in such a way that hands, cups, or other objects cannot contaminate the water.

**Water use:**
- Adequate amounts of water should be available and used for personal and domestic hygiene. (It is estimated that a minimum of 30–40 litres per person per day are needed for personal and domestic hygiene.)

**Food handling:**
- Hands should be washed with soap or ash before food is prepared or eaten.
- Vegetables and fruits should be washed with safe water, and food should be properly covered.
- Utensils used for food preparation and cooking should be washed with safe water as soon as possible after use and left in a clean place.

**Excreta disposal:**
- All men, women, and children should use latrines at home, at work, and at school.
- The stools of infants and young children should be safely disposed of.
- Household latrines should be sited in such a way that the pit contents cannot enter water sources or the groundwater table.
- Hand-washing facilities and soap or ash should be available, and hands should always be washed after defecation and after helping babies and small children.

**Wastewater disposal:**
- Household wastewater should be disposed of or reused properly. Measures should be taken to ensure that wastewater is not allowed to create breeding places for mosquitos and other disease vectors or to contaminate safe water.

## 7.2 Planning hygiene education

Planning hygiene education in a community involves the following steps:

- — dialogue with the community and local agencies;
- — selection of priority hygiene behaviours to be changed, based on surveillance data and felt needs within the community;
- — analysis of influence on selected behaviours and the implications for hygiene education.

Preparation of an *action plan* for hygiene education requires answers to the following questions:

- How will community participation be mobilized?
- Who should the education be directed at (target group)?
- What should the content of the education be?
- Who should carry out the hygiene education?
- What educational methods should be used?
- What support should be provided by the surveillance agency?

### 7.2.1 Community participation and empowerment

The importance of community participation has been stressed in earlier chapters. Hygiene behaviours are particularly difficult to change because they relate to daily activities, they are shared by the whole community, and they form part of the culture and traditions of the community. The improvement of water supply, sanitation, and hygiene should be seen as part of an overall process of community development. It is important, therefore, to work with the whole community and particularly with schoolchildren, and to involve them in all stages of hygiene education, including selecting priority hygiene behaviours, understanding the influences on such behaviours, selecting educational methods, and implementation. The educational methods used should be those that strengthen and empower individuals and communities to work for change.

There are no set rules for developing a community participation programme, but the stages described in Table 7.2 are common to many such programmes.

The community may already be highly organized and taking action on health issues. If so, only a few visits by surveillance field staff will be needed to introduce the concepts of surveillance and involve the community in the surveillance programme. However, it may be that there is no well developed structure, that sections of the community, such as women, are poorly represented, and that there are disagreements or factional conflicts. In this situation, achieving community participation will take more time and require many visits by field staff to bring people together, resolve differences, agree on common aims, and take action. Even after the community starts to become involved, further visits, possibly over several years, will be needed to provide support and encouragement, and ensure that the structures created continue to operate.

***Table 7.2 Stages in the community participation process***

**Getting to know the community:**
- learning about the community, its structure and leadership pattern
- initial contacts with families, leaders and community groups
- dialogue and discussion on concerns and felt needs

**Organization building:**
- strengthening of community organization
- establishment of new structures, e.g. water committees, women's groups
- educational activities within community structures
- decision-making on priorities
- selection of community members for training as water leaders

**Initial actions:**
- action by the community on achievable short-term goals that meet felt needs and bring the community together
- reflection on initial activities
- setting of priorities for future activities

**Further actions:**
- activities in which the community takes a greater share of responsibility for decision-making and management

## 7.2.2 Selection of behaviours to be changed

It is better to concentrate on a small number of behaviours than to attempt to influence all the hygiene behaviours listed in Table 7.1. The behaviours chosen should be selected on the basis of probable public health benefit to the community. Some of the questions that will need to be asked in order to determine priorities include the following:

- What is the evidence that the behaviour represents a problem in the community?
- Which behaviour changes will have the greatest impact on improving health?
- Which hygiene behaviours will be the easiest to change?
- What are the specific requirements of the water-supply and sanitation systems that are being promoted in the community?
- What are the felt needs and priorities of the community?

It is best to concentrate on those hygiene practices shown by the surveillance to be a priority for remedial action in the community concerned; these should be the practices which are likely to be of the greatest benefit to health. However, greater efforts will be required to change hygiene practices that the community does not see as important or that conflict with its culture and traditions.

### 7.2.3 Factors influencing hygiene behaviour and selection of content of education

Hygiene education programmes should be based on an understanding of the factors that influence behaviour at the community level. These might include:

- — enabling factors such as money, materials, and time to carry out the behaviour;
- — pressure from particular members of the family and community, e.g. elders, traditional healers, opinion leaders;
- — beliefs and attitudes among community members with respect to the hygiene behaviour, and especially the perceived benefits and disadvantages of taking action, and the understanding of the relationship between health and hygiene.

An understanding of the factors that influence hygiene behaviours will help in identifying the resources (e.g. soap, storage containers), the key individuals in the home and community, and the important beliefs that should be taken into account. This will help to ensure that the content of the hygiene education is relevant to the community. Good advice should:

- — result in improved health
- — be affordable
- — require a minimum of effort and time to put into practice
- — be realistic
- — be culturally acceptable
- — meet a felt need
- — be easy to understand.

One of the most important characteristics of effective health education is that it builds on concepts, ideas, and practices that people already have. Most communities already have beliefs about cleanliness, diarrhoea, and hygiene. In the short term, it may not be necessary to convince people of the correctness of the germ theory of disease in order to get them to use latrines and practise good hygiene. This is a long-term objective that is best achieved in schools. It is possible to find supporting ideas in many traditional belief systems, and to appeal, for example, to the desire for comfort and privacy.

### 7.2.4 Selection of target groups

Hygiene education is aimed at two kinds of target group:

- *Primary target group*—the members of the household, children, women, men, grandparents, and others who care for children.
- *Secondary target group*—people who need to be involved in the programme because of the influence that they have in the community (local leaders, field staff from other agencies, politicians, traditional healers, etc.).

A single hygiene education message and the associated materials are unlikely to be sufficient for all purposes. Ideally, the individual needs of each of the target groups in the community should be addressed, taking into account educational level and any cultural constraints.

### 7.2.5 Information needs for hygiene education

Before a formal hygiene education programme is begun, it is important to include in the sanitary survey (see Chapter 3) an assessment of the sociocultural factors that characterize the community, in order to determine:

- — local beliefs and attitudes regarding water, sanitation, and health;
- — traditional water use and defecation habits and excreta disposal practices;
- — current levels of knowledge about disease transmission, especially among community leaders and other influential individuals;
- — the priority given to improvements in water supply and sanitation in relation to other community needs;
- — existing channels of communication in the community including books, newspapers, and magazines, radio or television, traditional drama, songs, and story-telling;
- — the members of the community and field workers from other agencies who might be involved in hygiene education activities.

## 7.3 Educational methods

Some key characteristics of effective communication and health education are summarized in Table 7.3.

The choice of methods to be used should take account of the nature of what is to be conveyed and of local beliefs, values, and practices; the characteristics of the intended audience, including educational and literacy levels and exposure to,

***Table 7.3 Characteristics of effective health education***

- Promotes actions that are realistic and feasible within the constraints faced by the community
- Builds on ideas and concepts that people already have and on common practices
- Is repeated and reinforced over time using different methods
- Uses existing channels of communication, e.g. songs, drama, and story-telling, and can be appropriately adapted to these media
- Is entertaining and attracts the community's attention
- Uses clear simple language and local expressions, and emphasizes the short-term benefits of action
- Provides opportunities for dialogue and discussion to allow learner participation and feedback
- Uses demonstrations to show the benefits of adopting the practices recommended

and familiarity with, different educational methods; practical considerations, including the amount of money available and the experience of the staff.

Effecting the fundamental changes in lifestyle that are required to improve sanitation and hygiene will usually require an intensive programme of face-to-face communication in the community. This might include visiting individual householders or working with groups in community or other local settings: women's groups, mothers' groups, children in schools, or trade unions.

In hygiene education, it is important to emphasize *participatory learning methods*; these can include small-group teaching, simulations, case studies, group exercises, and role play. These methods:

- — avoid formal lecture presentations
- — encourage discussion between participants
- — encourage interaction during the sessions
- — use a variety of games, puzzles, and exercises
- — use learning aids that stimulate discussion and comments.

Participatory learning methods have a number of advantages: their active nature means that participants are more likely to remember what they have learned; participants draw from their own experience and are allowed to discover principles for themselves; opportunities are provided for learning problem-solving skills; participants acquire the confidence to tackle problems and improve their conditions. However, many field staff will be unfamiliar with participatory learning methods and will require training.

Traditional media such as drama, songs, and story-telling are of great potential value and have been used for education in sanitation and hygiene. They combine entertainment with practical advice and can be used to stimulate discussion and community participation. The actors and musicians should be given the basic information on hygiene and health, and allowed to produce a performance that is both entertaining and understood by the community. It can also be valuable to involve members of the community in the performance.

One of the most powerful forms of communication is through real-life examples, e.g. a demonstration latrine can be constructed in a well-chosen location, correct practices can be demonstrated. Demonstrations are most effective if they can be seen to produce observable benefits in the short term. However, since the health benefits of sanitation and hygiene can take time to become apparent, it is best to emphasize immediate benefits such as convenience, comfort, and freedom from flies and smells.

Valuable messages can also be communicated by "satisfied acceptors"—people who have improved their sanitation or hygiene practices and are pleased with the results. They are the best people to explain the benefits to others, as they will use everyday language and will have greater credibility with the community.

A range of learning materials such as flannelgraphs, flip-charts, leaflets, posters, slide presentations, videos, and models can be developed to support the work. These should be pretested on a sample of the intended audience to ensure that their messages are easily understood, and that the advice is relevant and meets the community's needs. Local artists can be used and encouraged to work with the community in preparing materials.

In general, health education messages should be reinforced by repetition, ideally through more than one medium.

Face-to-face education can be supported by the mass media such as radio, television, and newspapers if the initial survey shows that these will reach the community. Carefully designed and tested radio programmes, for example, can be used to spread simple information rapidly to large numbers of people, and to stimulate increasing awareness of, and interest in, the education programme. Broadcasts should use a variety of entertaining and interesting formats such as songs, dramas, quizzes, and interviews with members of the community. The timing of such broadcasts should fit in with community activities. Because the mass media reach large audiences, it is difficult to make messages specific to local communities; it may be useful to prepare radio programmes on cassettes, which can be played to small groups or through loudspeakers in public places.

A longer-term approach to improving hygiene in the community is working with children in schools. This enables the concepts of hygiene to become part of a general understanding of health and the influence of the environment. School-children can then introduce hygiene concepts to their parents. They learn from what they see around them, so that the school environment itself should meet the requirements of good hygiene, for example by providing latrines, water for hand-washing, generally clean surroundings, and hygienic facilities for the preparation and serving of school meals.

Hygiene education can take place in the classroom but also through activities in the school surroundings and community. It can be taught as part of a health education programme as well as of other subjects, such as mathematics, art, science, music, and drama, and should be integrated within a broad-based health-education programme with a limited number of predefined educational objectives focused on the health needs of the community. This should provide a basic knowledge of health in the first years of school that can be extended by a more detailed discussion of health and disease with older schoolchildren.

## 7.4 Human resources for hygiene education

For a hygiene education programme to be effectively implemented, management staff must be aware of its importance and committed to it in practice. Such staff include sanitary engineers and specialists in public health medicine, and hygiene education should form part of their professional training.

The effectiveness of hygiene education within surveillance programmes will depend on the extent to which local resources can be mobilized to support educational activities.

Most hygiene practices are established early in life and reinforced by parents and elders in the family. In particular, mothers play an important role in encouraging hygiene routines in their children and, in most communities, are involved in the organization of the home, the collection and storage of water, cleanliness, and child care. An essential priority in hygiene education is therefore to involve women, by working either with individual women in their homes or with women's groups within the community. Women should be represented in any community groups that are formed as part of the surveillance programme.

The most important resource for hygiene education is the community itself. A search should be made for any groups or organizations in the community that might be involved in hygiene education including village committees, water committees, health committees, young farmers' clubs, women's groups, and religious bodies.

Hygiene education is already part of the activities of many members of the community and field agencies (see Table 7.4), as well as of the staff of clinics and health centres. Community health workers in primary health care programmes are key health educators at the village level. Public health inspectors and rural health assistants are heavily involved in hygiene education as part of their promotion of safe water, environmental sanitation, and hygiene. Health workers in hospitals have a health education role as part of the treatment and rehabilitation process.

### *Table 7.4 Potential human resources for hygiene education in the community*

**Health services:**
Doctors and nurses in primary health care
Midwives
Health visitors
Public health nurses
Medical assistants
Nutrition programmes
Immunization programmes
Special disease programmes
Village health workers
Sanitary technicians
Veterinarians

**Public health services:**
Public health inspectors
Water supply staff
Sanitary technicians
Hygiene inspectors
Refuse management staff
Sanitary engineers

**Agricultural and development workers:**
Agricultural extension workers
Community development workers
Nutrition programme staff
Cooperative workers
Employment-generating programme staff
Women's programme staff

**Education services:**
Teachers in primary and secondary schools
Adult education staff
Literacy programme staff
Preschool programme staff
Vocational trainers

**Informal resources in the community:**
Elders
Parents
Traditional birth attendants
Traditional healers
Village leaders
Religious leaders

Outside the health services, those who may become involved in hygiene education include teachers in schools, adult education, and literacy programmes. In order to enable them to fulfil this role, the ministry of education or its equivalent should ensure that subjects such as the environment, hygiene and health are included in educational programmes, where appropriate.

Other workers in the community can also be mobilized for hygiene education. Agricultural extension workers who advise communities on growing crops can also provide education on health and nutrition. Community development officers engaged in promoting community organizations and cooperatives can play a key role in promoting community action on health issues.

In addition to government agencies, many voluntary bodies are actively involved in health education, including nutrition groups, family-planning associations, and the Red Cross and Red Crescent and other societies.

When potential resources for hygiene education are being sought, the following questions should be asked: Are any groups involved in hygiene education at present? How likely is it that they will support hygiene education? What support would they need to become involved in hygiene education, e.g. training, learning resources?

Field staff and volunteers may need training in hygiene education, particularly in the newer participatory learning methods. The aim should be to develop self-sustaining programmes of hygiene education as part of the normal workload of local fieldworkers in the community. Although initially such fieldworkers may need training, support, and encouragement to undertake hygiene education, with time they should be capable of doing so with minimal external support.

## 7.5 Role of the surveillance agency in hygiene education

Hygiene education is only one of the many responsibilities of surveillance field staff. Many agencies may play a role in hygiene education, including government bodies (e.g., ministries of water, the environment, health, education, rural development, and local government), nongovernmental organizations (both national and international), and private institutions. Typically, a government agency will play a coordinating role which, because of the intersectoral nature of the activity, may involve the following:

*At the national level*:

- Working with other agencies including health education services, water supply services, and NGOs, and involving them in hygiene education activities.
- Undertaking hygiene education through the mass media to support activities at the community level.
- Reviewing, analysing and interpreting surveillance data in order to evaluate hygiene education activities and determine priority areas for future action.

- Collecting information on innovative and effective methods of hygiene education including national and foreign experience, and disseminating it through reports, workshops and meetings.
- Providing regional training in hygiene education for surveillance field staff and support agencies.

*At the regional level:*
- Acting as a bridge between activities at a national level and those in the region, briefing regional officials in project areas, providing details of national activities, and mass media programmes.
- Working with regional agencies to involve field personnel from as wide a range as possible of agencies, e.g. health assistants and health inspectors, nurses, village health workers, teachers, agricultural and rural development staff, rural home-makers, adult literacy and adult education workers.
- Coordinating the activities of various field agencies involved in hygiene education including advising on the content of educational programmes to ensure that they are accurate, relevant, and appropriate to the needs of local communities.
- Providing training in sanitation and hygiene education, including practical communication skills.
- Distributing educational materials and working with field staff and the community to produce locally relevant educational materials.
- Working with other field agencies and the community to ensure that reports on surveillance activities include information on hygiene education needs, the effectiveness of local activities, and research on hygiene education.

*At the local level:*
- Working with families and communities to stimulate community participation and undertake hygiene education.
- Working with community organizations engaged in hygiene education and surveillance activities, e.g. water committees, to provide support and training, and involving them in hygiene education, monitoring, and surveillance activities.
- Working with field staff from different agencies active in the local communities, and coordinating hygiene education, training, support, and educational materials.

## 7.6 Funding hygiene education activities

Because of the intersectoral nature of hygiene education, a number of agencies will obviously make contributions both in financing and in kind. Thus, for example, the education sector may contribute significantly through schools and adult literacy or vocational training programmes, and the communities themselves may make significant contributions, especially in kind.

In practice, dedicated hygiene education programmes are most commonly the responsibility of the ministry of health or its equivalent. This is logical because of the responsibility of this agency for protecting health. Nevertheless, depending on local circumstances, other agencies can usefully link hygiene education activities to their programmes, e.g. mobile borehole drilling teams of the ministry responsible for water can be linked to hygiene educators.

# 8.

# Legislative, regulatory, policy, and basic management aspects

## 8.1 Application of water-supply legislation

### 8.1.1 Short- and medium-term targets

The *Guidelines for drinking-water quality* cover a large number of possible contaminants in order to meet the varied needs of countries. However, it is very unlikely that all of the contaminants mentioned will occur in a water supply. Care should therefore be taken in selecting substances for which national standards will be developed. A number of factors should be considered, including the geology of the region and the types of human activities that take place there. Thus, if a particular pesticide is not used in the region, there will be no need to monitor it or to establish a drinking-water standard for it. Scarce resources should not be wasted on developing standards for, and monitoring, substances of minor importance.

In countries where economic and human resources are limited, short- and medium-term targets should be set in establishing national drinking-water standards, water-quality surveillance, and quality-control programmes so that the most significant risks to human health are controlled first. It is thus important to draft water-quality legislation in such a manner as to allow for flexibility in achieving water-quality targets in stages.

The most common and widespread health risk associated with drinking-water is its microbial contamination, the consequences of which are so serious that its control must always be of paramount importance. Microbiological quality should therefore be regarded as a priority, although it may be impossible to attain the targets in the short or medium term. It is therefore necessary to ensure that priority is given to water supplies presenting the greatest public health risk, whether through prioritization, as described in Chapter 5, or through legal mechanisms such as exemptions to allow for progressive improvements.

Attempting to follow the *Guidelines* in an indiscriminate manner can result in a situation where the drinking-water standards adopted in a country are not appropriate to its real health needs, or where there is little or no professional or economic capability to monitor and enforce them. In such a situation, personnel concerned with water quality and public health and community leaders may become demoralized, leading to a loss of confidence in all water-quality standards and monitoring procedures, frustration, and loss of respect for health and environmental laws and regulations in general.

Health and water authorities should therefore formulate a clear strategy for the establishment of water-quality goals in stages—short-, medium- and long-term. A programme based on modest but realistic goals including fewer water-quality parameters but at attainable levels consistent with providing a reasonable degree of public health protection may achieve more than an overambitious one.

The drinking-water quality legislation should clearly provide for the possibility of regional differences in standards and for differences between large urban and small-community supplies. This can also take the form of temporary exemptions for certain communities or areas from specified water-quality standards for clearly defined periods of time. Such exemptions should be granted by a senior public or environmental health official at the district, regional, or national level who has authority under the law to do so.

Interim standards, permitted deviations and exemptions should be established under the authority of the law as part of a national or regional policy, rather than as a result of local initiatives and self-interest. Water-supply agencies should act on all matters relating to the quality of the water that they supply under the authority of laws and regulations laid down by a higher authority, rather than by establishing their own interim standards based on their own judgement or convenience. Such a legal framework is important both for ensuring public health protection and to protect water-supply agencies from being held liable for "substandard" water.

### 8.1.2 Compliance: the role of the water-supply agency and the surveillance agency

Legislation should clearly specify that the *water-supply agency* is legally responsible at all times for the quality of the water sold and/or supplied to the consumer and for the proper supervision, inspection, maintenance, and safe operation of the water-supply system. It is the water-supply agency which actually provides water to the public—the "customer"—usually on a commercial basis, and which should, as the supplier or vendor of the finished product, be legally responsible (under both criminal and civil law) for its quality and safety from a public health point of view. However, it should be held responsible for the quality of the water only up to a defined point in the distribution system, and not for any deterioration of water quality within the household as a result of poor plumbing or unsatisfactory storage tanks. It should be the long-term policy of the health and water authorities to place the legal burden of the primary level of water-quality control testing on the supply agency. These authorities should develop the infrastructure necessary for quality control, the costs involved forming part of the price of the water. This form of transfer of responsibility, in a decentralized manner, to the producer/supplier provides a system of independent surveillance coupled with strict enforcement by an authority with the power to determine whether the supplier is fulfilling its responsibilities, and is based on the principles of sound administration.

The legislation should empower the appointed *surveillance agency* to enforce compliance with water-quality standards and regulations by carrying out independent periodic surveillance of all aspects of water quality and safety, including sanitary inspections and spot checks. The results of this surveillance should be reported to the water-supply agency, which should be required to take remedial action, where necessary.

Surveillance should primarily be a support and advisory function and only secondarily one of enforcement and the imposition of penalties. However, appropriate penalties should be specified in the law, including fines for violations, and continuing fines for continuing violations. Consideration should be given to holding water-supply agency management personally responsible for serious offences involving personal neglect and mismanagement, something that has been found to be effective in certain countries. The surveillance agency should be required by law to publish annual reports on its work or at least to provide free public access to all water-quality surveillance results in a form that is both meaningful and comprehensible to the general public.

While remedial action to ensure the timely correction of faults should be an aim of the surveillance programme, there may at times be a need for penalties to ensure compliance. The surveillance agency must be supported by strong and enforceable legislation if it is to be effective. However, it is important that the agency develops a positive and supportive relationship with suppliers.

The surveillance agency should be empowered by law to compel the supplier to post notices recommending the boiling of water when microbial contamination is detected.

### 8.1.3 Surveillance requirements

The legislation should define the duties, obligations, and powers of the water-surveillance agency. Legal and organizational arrangements aimed at ensuring compliance with the legislation, standards, or codes of practice for drinking-water quality must provide for the establishment, wherever feasible, of an independent surveillance agency. Often the optimum procedure is to empower a government agency, as discussed in Chapter 2, which has qualified professional personnel and laboratory facilities to undertake the role of a surveillance agency. In many developing countries, however, the ministry of health or other surveillance agency may have the necessary power under the law but few resources for surveillance activities, and is thus ineffective. The delegation of the surveillance function to a qualified, government-authorized agency, possibly at a lower level (e.g. provincial or local) or to a private institution, may be considered as an alternative approach.

### 8.1.4 Sampling frequencies and parameters

Frequent sanitary inspections and water-quality testing, particularly for microbiological contamination, are essential elements in any surveillance programme aimed at ensuring that drinking-water meets the standards established. In rural areas, where water sources may not be exposed to industrial wastes or agricultural chemicals, testing for most micropollutants may not be necessary or feasible. Realistic and flexible sampling frequencies should be established for the parameters that are to be measured. The basic water legislation should not specify sampling frequencies but should give the administration the power to establish a list of parameters to be measured and the frequency of such measurements. However, it must be emphasized that water-supply surveillance is not based solely on laboratory testing, but also on regular sanitary inspections and surveys accompanied by recommendations for remedial action. Follow-up visits will also be required to ensure that such remedial action is taken.

### 8.1.5 Prescribed analytical methods

Drinking-water standards or regulations must be designed to ensure that accepted, standardized, reliable, and accurate analytical methods are used by all agencies and laboratories. This is particularly important where litigation may be necessary. Simpler and less expensive methods may sometimes be accepted for some routine tests in remote areas where there are no proper laboratory facilities. The legislation must allow for such alternative methods under certain circumstances. Regulations should also require quality-assurance procedures to be introduced and monitored in water-supply agencies and certified private laboratories carrying out water-quality testing.

## 8.2 Technical regulations: construction, operation, and plumbing codes of practice

Important elements in ensuring the supply of water of the required quality include proper source selection, and the design, construction, and operation of water-supply facilities. Codes of practice should be established to ensure that the best sustainable source of water is selected, and that systems are designed to protect water quality by means of effective barriers to contamination. These matters can be covered to some extent by technical regulations and statutory codes under the basic water-quality legislation. However, excessively rigid construction and plumbing codes which can only be amended by complicated and slow legislative procedures should be avoided. Such technical regulations and codes should be administrative in character and easy to amend to allow for new technological developments and for the introduction of low-cost interim methods under certain circumstances.

# Selected further reading


Bartram J. *Drinking water supply surveillance.* Guildford, England, Robens Institute of Health and Safety, University of Surrey, 1990.

*Disinfection of rural and small-community water supplies: a manual for design and operation.* Medmenham, England, Water Research Centre, 1989.

Foster S, Venura M, Hirata R. *Groundwater pollution.* Lima, WHO/PAHO/CEPIS, 1987.

*Guidelines for drinking-water quality. Vol. 1, Recommendations.* Geneva, World Health Organization, 1993.

*Guidelines for drinking-water quality. Vol. 2, Health criteria and other supporting information.* Geneva, World Health Organization, 1996.

Lloyd B, Helmer R. *Surveillance of drinking water quality in rural areas.* Harlow, England, Longman Scientific and Technical, 1991.

*Manual of basic techniques for a health laboratory.* Geneva, World Health Organization, 1980 (second edition in preparation).

Rajagopalan S, Shiffman MA. *Guide to simple sanitary measures for the control of enteric diseases.* Geneva, World Health Organization, 1974.

*Selected physical and chemical standard methods for students.* Washington, DC, American Public Health Association, 1990.

*Small community water supplies—technology of small water supply systems in developing countries.* The Hague, International Reference Centre for Community Water Supply and Sanitation, 1983.

*Standard methods for the examination of water and wastewater,* 17th ed. Washington, DC, American Public Health Association, 1989.

*Surveillance of drinking-water quality.* Geneva, World Health Organization, 1976 (WHO Monograph Series, No. 63).

**International Organization for Standardization (ISO) standards for sampling of drinking-water supplies:**

| | | |
|---|---|---|
| ISO 5667-1:1980 | Sampling—Part 1 | Guidance on the design of sampling programmes |

| | | |
|---|---|---|
| ISO 5667-2:1991 | Sampling—Part 2 | Guidance on sampling techniques |
| ISO 5667–3:1994 | Sampling—Part 3 | Guidance on the preservation and handling of samples |
| ISO 5667-4:1987 | Sampling—Part 4 | Guidance on sampling from lakes, natural and man-made |
| ISO 5667-5:1991 | Sampling—Part 5 | Guidance on sampling of drinking-water and water used for food and beverage processing |
| ISO 5667-6:1990 | Sampling—Part 6 | Guidance on sampling of rivers and streams |

**International Organization for Standardization (ISO) standards for microbiological analysis:**

| | |
|---|---|
| ISO 6222:1988 | Enumeration of viable micro-organisms. Colony count by inoculation in or on a nutrient agar culture medium. |
| ISO 7899-1:1984 | Detection and enumeration of faecal streptococci—Part 1: Method by enrichment in a liquid medium. |
| ISO 7899-2:1984 | Detection and enumeration of faecal streptococci—Part 2: Method by membrane filtration. |
| ISO 8199:1988 | General guide to the enumeration of micro-organisms by culture. |
| ISO 9308-1:1990 | Detection and enumeration of coliform organisms, thermotolerant coliform organisms and presumptive *Escherichia coli*—Part 1: Membrane filtration method. |
| ISO 9308-2:1990 | Detection and enumeration of coliform organisms, thermotolerant coliform organisms and presumptive *Escherichia coli*—Part 2: Multiple tube (most probable number) method. |

Annex 1

# List of contributors and reviewers

H. Abouzaid, Environmental Health Risk Assessment and Management, Division of Environmental Health, WHO Regional Office for the Eastern Mediterranean, Alexandria, Egypt

J. Bartram, Manager, Water and Wastes, WHO European Centre for Environment and Health, Rome, Italy

V. Bizhga, Cabinet of the Minister, Ministry of Health and Environmental Protection, Tirana, Albania

X. Bonnefoy, Environmental Health Planning/Ecology, World Health Organization, Regional Office for Europe, Copenhagen, Denmark

T. Boonyakarnkul, Environmental Health Division, Health Department, Ministry of Public Health, Bangkok, Thailand

M.T. Boot, Programme Officer, IRC International Water and Sanitation Centre, The Hague, Netherlands

J.Z. Boutros, Consultant in Food and Water Control, Khartoum, Sudan

B. Clarke, Centre for Environmental Health and Water Engineering, Department of Civil Engineering, University of Surrey, Guildford, England

W. Fellows, Programme Officer, Water and Environmental Sanitation, UNICEF, Harare, Zimbabwe

H. Fenger, WHO Regional Office for Europe, Copenhagen Denmark

H. Galal-Gorchev, International Programme on Chemical Safety, World Health Organization, Geneva, Switzerland

E. Gonzalez, Chief, Division of Water Quality, Water Supply and Sewerage, San José, Costa Rica

F.J. Gumbo, Head of Water Laboratories, Operation, Maintenance and Water Laboratories Division, Ministry of Water (MAJI), Dar-es-Salaam, United Republic of Tanzania

A.N. Havelaar, National Institute of Public Health and Environmental Protection, Bilthoven, Netherlands

R. Helmer, Urban Environmental Health, Division of Operational Support in Environmental Health, World Health Organization, Geneva, Switzerland

G. Howard, Research Fellow, The Robens Institute of Health and Safety, University of Surrey, Guildford, England

J. Hubley, Senior Lecturer in Health Education, Health Education Unit, Faculty of Health and Social Care, Leeds Metropolitan University, Leeds, England

J.A. Hueb, Rural Environmental Health, Division of Operational Support in Environmental Health, World Health Organization, Geneva, Switzerland

B. Jackson, Senior Engineering Advisor, British Development Division in East Africa, Nairobi, Kenya; Water Resources Adviser, Engineering Division, Overseas Development Administration, London, England

E. Khaka, Ministry of Lands, Agriculture and Water, Harare, Zimbabwe

L. Laugeri, formerly Community Water Supply, Division of Environmental Health, World Health Organization, Geneva, Switzerland

S. Laver, Lecturer, Department of Community Medicine, University of Zimbabwe, Mount Pleasant, Harare, Zimbabwe

L. Leno, Microbiologist, Research Institute of Hygiene and Epidemiology, and Cabinet of the Minister, Ministry of Health and Environmental Protection, Tirana, Albania

B. Lloyd, Head, Centre for Environmental Health and Water Engineering, Department of Civil Engineering, University of Surrey, Guildford, England

M.T. Martins, Associate Professor, Environmental Microbiology Laboratory, University of São Paulo, Brazil

L. Monjour, Water, Agriculture and Health in the Tropics, Paris, France

P. Morgan, Advisor, Water and Sanitation, Ministry of Health, Blair Research Laboratory, Harare, Zimbabwe

S. Mtero, Principal Medical Research Officer, Ministry of Health, Blair Research Laboratory, Harare, Zimbabwe

S. Musingarabwi, Director, Environmental Health Services, Ministry of Health, Harare, Zimbabwe

F. Niang, Chief, Laboratory Service, Senegalese National Water Management Company, Dakar, Senegal

M. Pardon, DelAgua/CEPIS, Lima, Peru

S. Pedley, Research Fellow, The Robens Institute of Health and Safety, University of Surrey, Guildford, England

E.B. Pike, formerly Water Research Centre, Medmenham, England

J. Ptashekas, Director, Environmental Medicine Centre, Vilnius, Lithuania

P. Ranque, Dracunculiasis Eradication, Division of Control of Tropical Diseases, World Health Organization, Geneva, Switzerland

P.K. Ray, Director, Industrial Toxicology Research Centre, Lucknow, India

A. Rickards, Research Fellow, The Robens Institute of Health and Safety, University of Surrey, England

K. Shehu, Director, Hydraulics, Water Supply and Sewerage, Tirana, Albania

H. Shuval, Lunenfield-Kunin Professor of Environmental Health, Hebrew University of Jerusalem, Israel

F. Solsona, WHO Sanitary Engineer, Brasilia, Brazil

P. Taylor, Director, Training Centre for Water and Sanitation, Department of Civil Engineering, University of Zimbabwe, Harare, Zimbabwe

J. Thoegersen, Centre for Developing Countries, Technical University of Denmark, Lyngby, Denmark

H. Utkilen, Scientist, National Institute of Public Health, Department of Environmental Medicine, Oslo, Norway

K. Wedgwood, Research Officer, The Robens Institute of Health and Safety, University of Surrey, Guildford, England

F. Zawide, WHO Sanitary Engineer, Sub-region III, Harare, Zimbabwe

Annex 2

# Examples of sanitary inspection forms

Examples of sanitary inspection forms are given here as follows:

Nonpiped supplies: open dug well (Fig. A2.1); dug well with windlass and partial cover (Fig. A2.2); covered dug well with hand-pump (Fig. A2.3); rain-water collection and storage (Fig. A2.4); tubewell with hand-pump (Fig. A2.5); tanker trucks, filling stations, and household tanks (Fig. A2.6).

Piped supplies: deep borehole with mechanical pump (Fig. A2.7); protected spring source (Fig. A2.8); surface sources and abstraction (Fig. A2.9); piped distribution (Fig. A2.10); water-treatment plant (Fig. A2.11).

With the exception of Fig. A2.11, these all consist of two pages and include illustrations depicting the various types of water supplies in appropriate settings; potential hazards are listed and numbered. The use of these forms enables a hazard score to be assigned to the particular water supply based on the total number of hazards found; however, differential weighting may be necessary to allow for local conditions (see p. 47).

Latrines and other point sources of potential faecal contamination should be located sufficiently far from groundwater sources used for drinking purposes to ensure that the risk of pathogen survival is very low. Once the "travel time" of microbial pathogens through the ground has been established, it is possible to determine a minimum safe distance (MSD) of potentially polluting activities from water sources.

The travel time of microbes depends on local hydrogeological conditions, in particular the hydraulic conductivity or permeability of the soil and rock in the unsaturated and saturated zones. Thus it is difficult to set MSDs that are universally applicable. Travel time will also be affected by the volume and concentration of contamination introduced into an area. It has been shown that in rural areas of low population density, the majority of viruses and bacteria will die after 10 days in groundwater. Thus, in these areas, where small-scale water supplies and sanitation are introduced, this travel time may be used as a basis for establishing MSDs. In urban areas where municipal wastewater is discharged and in those where there is intensive use of on-site sanitation, a figure of 50 days is more appropriate.

It is often difficult to obtain sound hydrogeological information. However, some idea of the local hydrogeological conditions can be gained by carefully recording the changes in soil and rock type during test drilling and by conducting infiltration tests in the area where latrine construction is proposed. The infiltration capacity of the soil in the area should be assessed when the water table is at its highest.

An infiltration test can be carried out as follows:

- Bore a hole(s) of diameter 10 cm and depth slightly greater than the maximum depth of the latrine pits (usually about 3 m).
- Fill the hole(s) with water and leave overnight to allow the soil to become saturated. When the soil is very dry, it may be necessary to refill the hole several times to ensure that the surrounding soil becomes saturated.
- Refill the hole(s) to a depth of 30 cm, and measure the fall in water level over 30- and 90-minute periods. The infiltration rate can then be estimated from the fall in water level during these periods. For greater accuracy, the volume of water infiltrating should be calculated and a value of the infiltration rate obtained in $m^3/m^2$ per hour or m/h.

It should be noted that the above test gives the infiltration capacity of the soil, i.e. the steady-state capacity to absorb water. Under very dry conditions, the actual infiltration rate may vary considerably. The test will usually be carried out with "clean" water, but the liquid from pit latrines will be "dirty" and the true infiltration capacity may therefore be lower. However, it is always better to be careful when locating latrines, and using clean water is likely to give a MSD which will be more than adequate for "dirty" water.

The procedure described above is a basic test which only gives an idea of how quickly liquid from a pit latrine will infiltrate and move through the ground. For greater precision, the hydraulic conductivity of the soil should be established by means of more sophisticated formulae based on Darcy's law, for which reference should be made to standard texts on groundwater and hydrogeology. Information should be obtained on the geology of the area where infiltration capacity is being evaluated, particularly on whether any fissures or joints underlie the area proposed for latrine development, since these may dramatically increase the hydraulic conductivity and thus the MSD.

The rate of movement of groundwater varies greatly depending on the permeability, ranging from fractions of metres per day in clays, to 1–10 m per day in sands, 50 m plus per day in very permeable gravels, and even greater rates in rock fissures, e.g. in limestone. Thus, while the MSD for impermeable clays may be as low as a few metres, for sands this may increase to 100 m; in permeable gravel beds or areas where there are shallow aquifers in fissures, it may reach as much as several kilometres.

As a rough guide, a value of 10 m can be considered as the absolute MSD allowable in areas of deep impermeable clay which does not form cracks during dry periods. However, unless detailed investigations of the area have been carried out under all conditions, it is preferable to increase this distance to at least 30 m. If the groundwater in the area is found in very permeable aquifers, such as gravels and rock fissures, on-site sanitation may not be appropriate. If no other option is available, sealed pits with impermeable concrete linings should be used.

### *Fig. A2.1 Example of sanitary inspection form for open dug well*

*Note*: MSD = minimum safe distance as determined locally; see section 6.2.2.

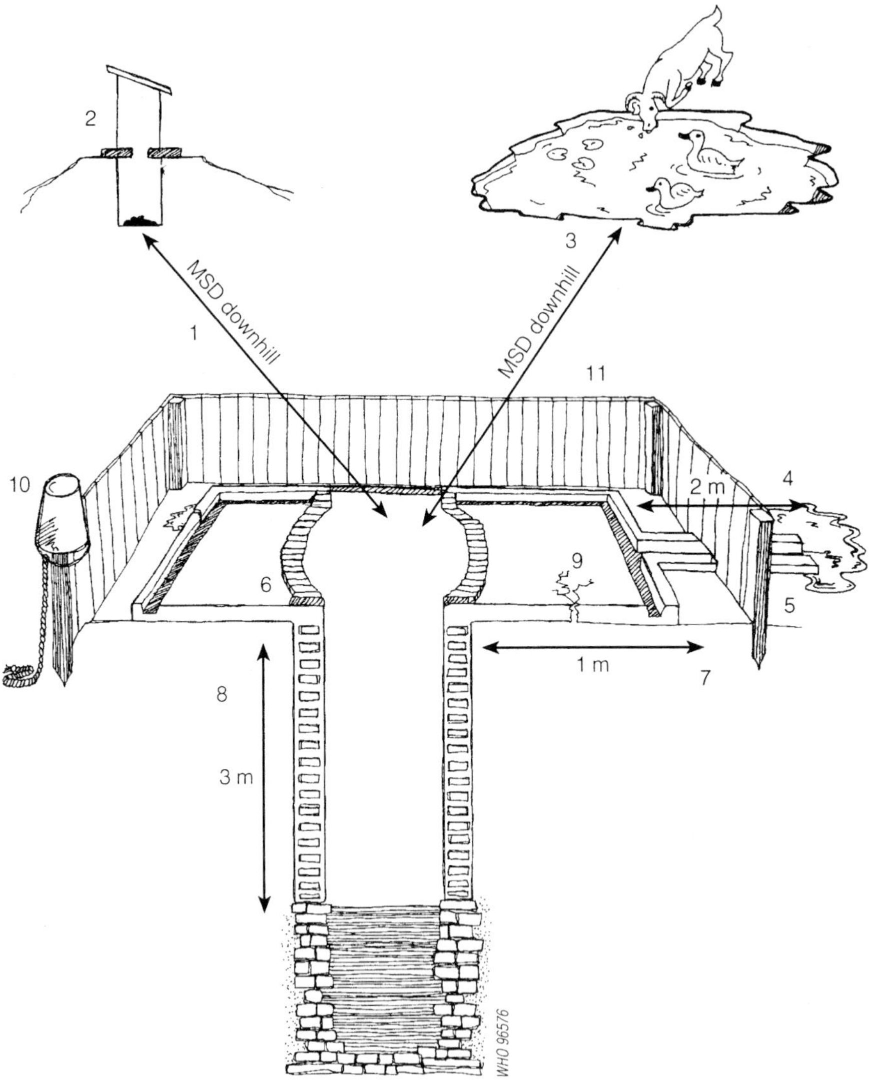

**I Type of facility OPEN DUG WELL**

1. General information: Health centre ..........................................................
   Village ..........................................................................
2. Code no.—Address ..............................................................................
3. Water authority/community representative signature .........................
4. Date of visit .....................................
5. Water sample taken? ....... Sample no. ......... Thermotolerant coliform grade .........

| **II Specific diagnostic information for assessment** | **Risk** |
|---|---|
| 1. Is there a latrine within 10 m of the well? | Y/N |
| 2. Is the nearest latrine on higher ground than the well? | Y/N |
| 3. Is there any other source of pollution (e.g. animal excreta, rubbish) within 10 m of the well? | Y/N |
| 4. Is the drainage poor, causing stagnant water within 2 m of the well? | Y/N |
| 5. Is there a faulty drainage channel? Is it broken, permitting ponding? | Y/N |
| 6. Is the wall (parapet) around the well inadequate, allowing surface water to enter the well? | Y/N |
| 7. Is the concrete floor less than 1 m wide around the well? | Y/N |
| 8. Are the walls of the well inadequately sealed at any point for 3 m below ground? | Y/N |
| 9. Are there any cracks in the concrete floor around the well which could permit water to enter the well? | Y/N |
| 10. Are the rope and bucket left in such a position that they may become contaminated? | Y/N |
| 11. Does the installation require fencing? | Y/N |
| Total score of risks ..................... | /11 |

Contamination risk score: 9–11 = very high; 6–8 = high; 3–5 = intermediate;
0–2 = low

**III Results and recommendations**

The following important points of risk were noted: ................................. (list nos 1–11) and the authority advised on remedial action.

Signature of sanitarian ........................................

### *Fig. A2.2 Example of sanitary inspection form for dug well with windlass and partial cover*

*Note*: MSD = minimum safe distance determined locally; see section 6.2.2.

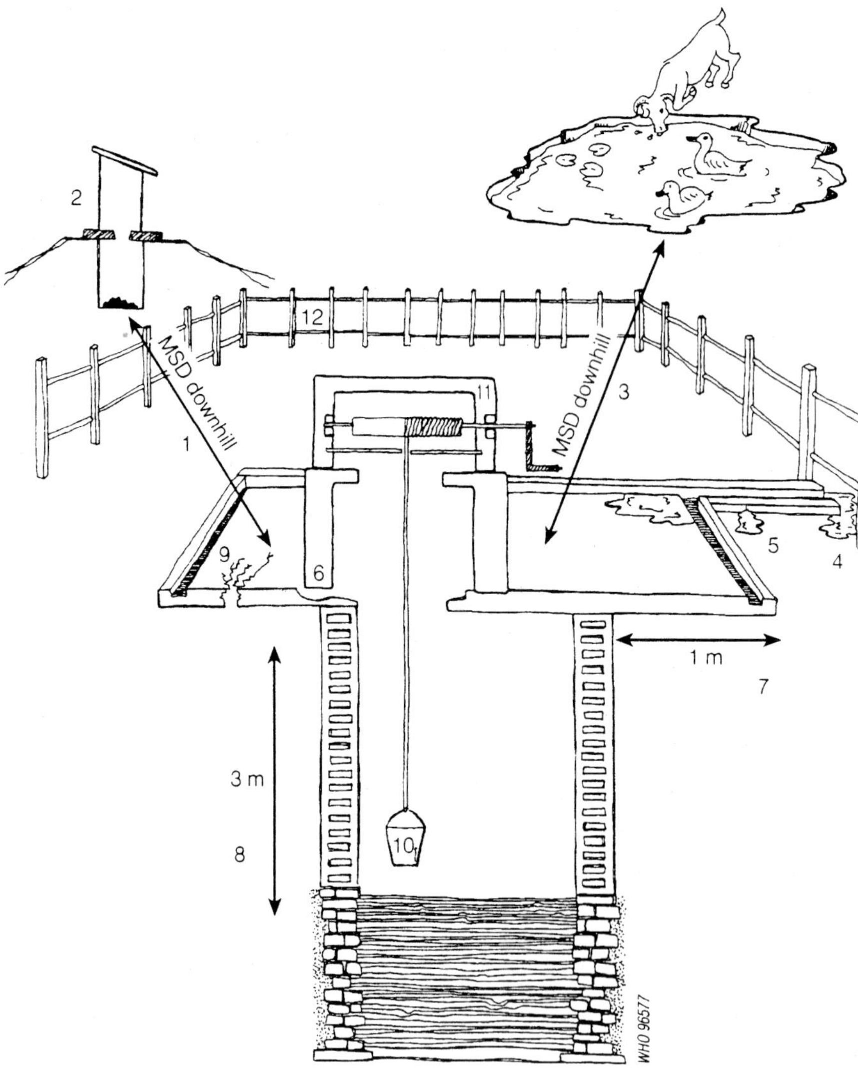

## I Type of facility DUG WELL WITH WINDLASS AND PARTIAL COVER

1. General information: Health centre ..........................................................
   Village ..........................................................................
2. Code no.—Address ...............................................................................
3. Water authority/community representative signature .........................
4. Date of visit .....................................
5. Water sample taken?....... Sample no......... Thermotolerant coliform grade .........

## II Specific diagnostic information for assessment

| | | Risk |
|---|---|---|
| 1. | Is there a latrine within 10 m of the well? | Y/N |
| 2. | Is the nearest latrine on higher ground than the well? | Y/N |
| 3. | Is there any other source of pollution (e.g. animal excreta, rubbish) within 10 m of the well? | Y/N |
| 4. | Is the drainage poor, causing stagnant water within 2 m of the well? | Y/N |
| 5. | Is there a faulty drainage channel? Is it broken, permitting ponding? | Y/N |
| 6. | Is the wall (parapet) around the well inadequate, allowing surface water to enter the well? | Y/N |
| 7. | Is the concrete floor less than 1 m wide around the well? | Y/N |
| 8. | Are the walls of the well inadequately sealed at any point for 3 m below ground? | Y/N |
| 9. | Are there any cracks in the concrete floor around the well which could permit water to enter the well? | Y/N |
| 10. | Are the rope and bucket left in such a position that they may become contaminated? | Y/N |
| 11. | Does the well require a cover? | Y/N |
| 12. | Does the installation require fencing? | Y/N |
| | Total score of risks .................... | /12 |

Contamination risk score: 9–12 = very high; 6–8 = high; 3–5 = intermediate; 0–2 = low

## III Results and recommendations

The following important points of risk were noted: ................................ (list nos 1–12) and the authority advised on remedial action.

Signature of sanitarian .......................................

## *Fig. A2.3 Example of sanitary inspection form for covered dug well with hand-pump*

*Note*: MSD = minimum safe distance determined locally; see section 6.2.2.

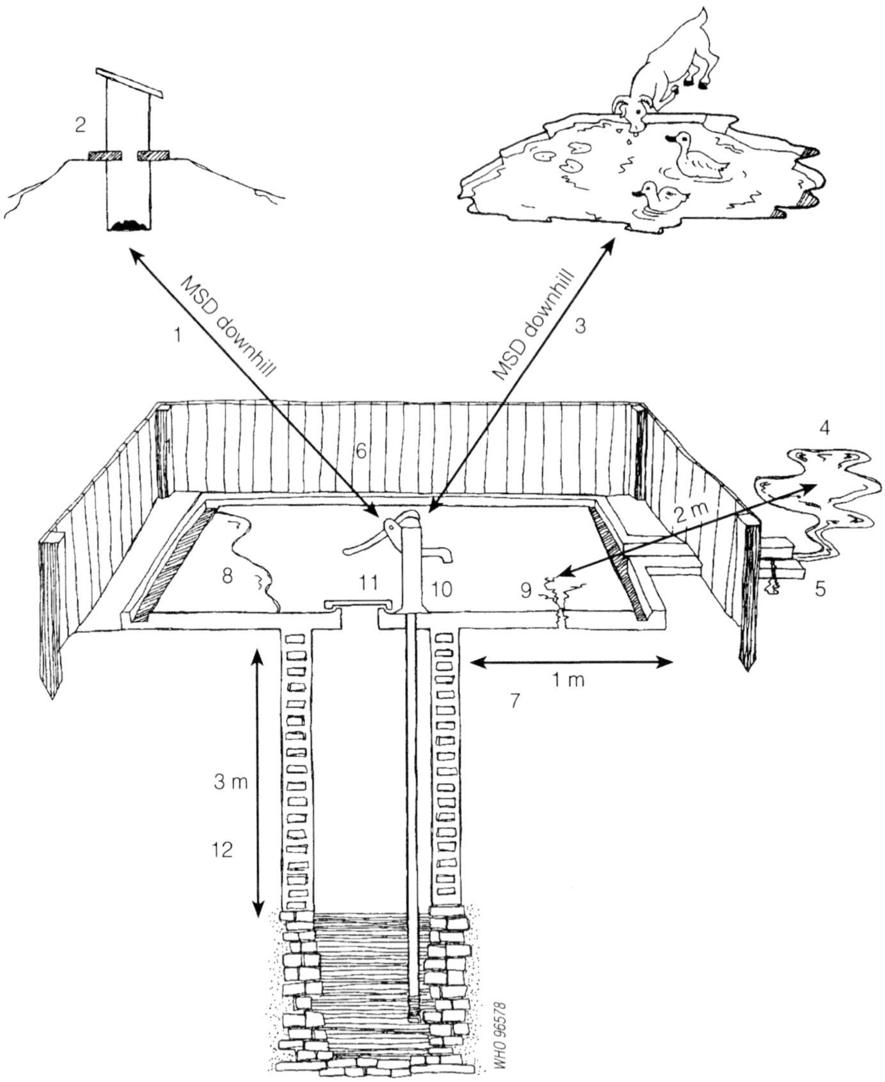

**I Type of facility** COVERED DUG WELL WITH HAND-PUMP

1. General information: Health centre ..................................................................
   Village ..............................................................................
2. Code no.—Address ..............................................................................
3. Water authority/community representative signature .........................
4. Date of visit ......................................
5. Water sample taken? ....... Sample no. ......... Thermotolerant coliform grade .........

| **II Specific diagnostic information for assessment** | **Risk** |
|---|---|
| 1. Is there a latrine within 10 m of the well and hand-pump? | Y/N |
| 2. Is the nearest latrine on higher ground than the hand-pump? | Y/N |
| 3. Is there any other source of pollution (e.g. animal excreta, rubbish) within 10 m of the hand-pump? | Y/N |
| 4. Is the drainage poor, causing stagnant water within 2 m of the cement floor of the hand-pump? | Y/N |
| 5. Is there a faulty drainage channel? Is it broken, permitting ponding? | Y/N |
| 6. Is the wall or fencing around the hand-pump inadequate, allowing animals in? | Y/N |
| 7. Is the concrete floor less than 1 m wide all around the hand-pump? | Y/N |
| 8. Is there any ponding on the concrete floor around the hand-pump? | Y/N |
| 9. Are there any cracks in the concrete floor around the hand-pump which could permit water to enter the hand-pump? | Y/N |
| 10. Is the hand-pump loose at the point of attachment to the base so that water could enter the casing? | Y/N |
| 11. Is the cover of the well unsanitary? | Y/N |
| 12. Are the walls of the well inadequately sealed at any point for 3 m below ground level? | Y/N |

Total score of risks .................... /12

Contamination risk score: 9–12 = very high; 6–8 = high; 3–5 = intermediate; 0–2 = low

**III Results and recommendations**

The following important points of risk were noted: ................................ (list nos 1–12) and the authority advised on remedial action.

Signature of sanitarian ........................................

***Fig. A2.4 Example of sanitary inspection form for rainwater collection and storage***

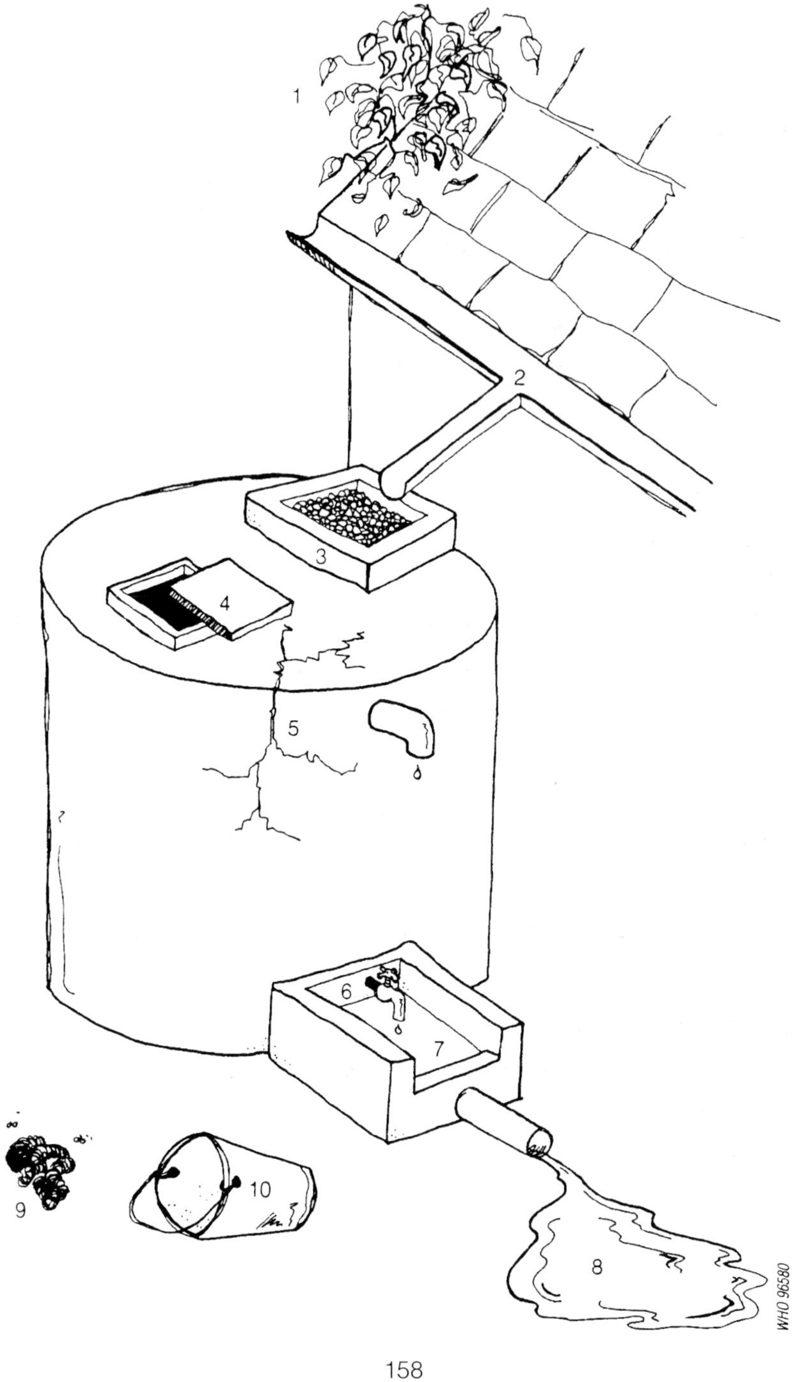

**I Type of facility** RAINWATER COLLECTION AND STORAGE

1. General information: Health centre ..........................................................................
   Village ..........................................................................
2. Code no.—Address ..........................................................................................
3. Water authority/community representative signature .........................
4. Date of visit .....................................
5. Water sample taken?....... Sample no.......... Thermotolerant coliform grade .........

**II Specific diagnostic information for assessment** **Risk**

1. Is there any visible contamination of the roof catchment area (plants, dirt, or excreta)? Y/N
2. Are the guttering channels that collect water dirty? Y/N
3. Is there any deficiency in the filter box at the tank inlet (e.g. lacks fine gravel)? Y/N
4. Is there any other point of entry to the tank that is not properly covered? Y/N
5. Is there any defect in the walls or top of the tank (e.g. cracks) that could let water in? Y/N
6. Is the tap leaking or otherwise defective? Y/N
7. Is the concrete floor under the tap defective or dirty? Y/N
8. Is the water collection area inadequately drained? Y/N
9. Is there any source of pollution around the tank or water collection area (e.g. excreta)? Y/N
10. Is a bucket in use and left in a place where it may become contaminated? Y/N

Total score of risks .................... /10

Contamination risk score: 9–10 = very high; 6–8 = high; 3–5 = intermediate;
0–2 = low

**III Results and recommendations**

The following important points of risk were noted: ................................ (list nos 1–10) and the authority advised on remedial action.

Signature of sanitarian ........................................

### *Fig. A2.5 Example of sanitary inspection form for tubewell with hand-pump*

*Note*: MSD = minimum safe distance determined locally; see section 6.2.2.

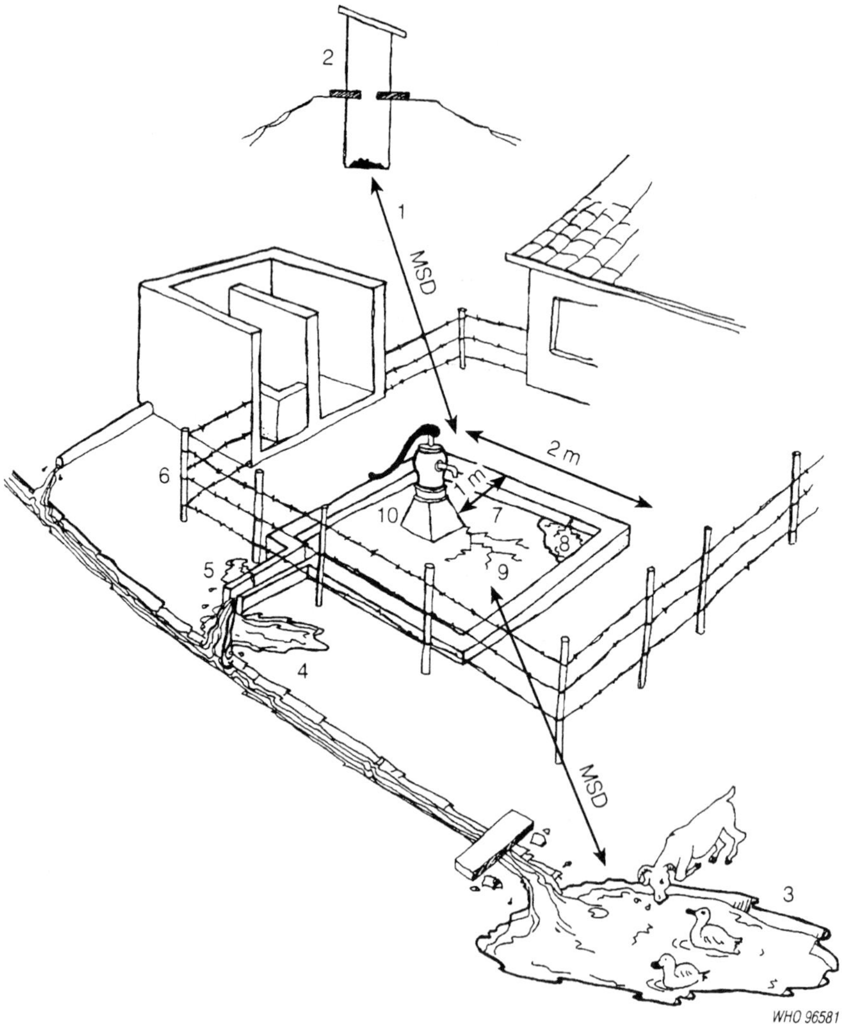

**I Type of facility** TUBEWELL WITH HAND-PUMP

1. General information: Health centre ....................................................................
   Village ....................................................................................
2. Code no.—Address ..................................................................................................
3. Water authority/community representative signature .........................
4. Date of visit .....................................
5. Water sample taken? ....... Sample no. ......... Thermotolerant coliform grade .........

| **II Specific diagnostic information for assessment** | **Risk** |
|---|---|
| 1. Is there a latrine within 10 m of the hand-pump? | Y/N |
| 2. Is the nearest latrine on higher ground than the hand-pump? | Y/N |
| 3. Is there any other source of pollution (e.g. animal excreta, rubbish, surface water) within 10 m of the hand-pump? | Y/N |
| 4. Is the drainage poor, causing stagnant water within 2 m of the hand-pump? | Y/N |
| 5. Is the hand-pump drainage channel faulty? Is it broken, permitting ponding? Does it need cleaning? | Y/N |
| 6. Is the fencing around the hand-pump inadequate, allowing animals in? | Y/N |
| 7. Is the concrete floor less than 1 m wide all around the hand-pump? | Y/N |
| 8. Is there any ponding on the concrete floor around the hand-pump? | Y/N |
| 9. Are there any cracks in the concrete floor around the hand-pump which could permit water to enter the well? | Y/N |
| 10. Is the hand-pump loose at the point of attachment to the base so that water could enter the casing? | Y/N |

Total score of risks .................... /10

Contamination risk score: 9–10 = very high; 6–8 = high; 3–5 = intermediate; 0–2 = low

**III Results and recommendations**

The following important points of risk were noted: ................................ (list nos 1–10) and the authority advised on remedial action.

Signature of sanitarian ........................................

***Fig. A2.6 Example of sanitary inspection form for filling stations, tanker trucks, and household tanks***

WHO 96582

WHO 96582a

**I Type of facility FILLING STATIONS, TANKER TRUCKS, AND HOUSEHOLD TANKS**

1. General information: Health centre ..........................................................................
   Village ..........................................................................
2. Code no.—Address ..........................................................................
3. Water authority/community representative signature ..........................
4. Date of visit ......................................
5. Water sample taken? ....... Sample no. ......... Thermotolerant coliform grade .........

| **II Specific diagnostic information for assessment** | **Risk** |
|---|---|
| *Tanker filling stations* | |
| 1. Is the chlorine level at the filling station less than 0.5 mg/litre? | Y/N |
| 2. Is the filling station excluded from the routine quality-control programme of the water authority? | Y/N |
| 3. Is the discharge pipe unsanitary? | Y/N |
| *Tanker trucks* | |
| 4. Is the tanker ever used for transporting other liquids besides drinking-water? | Y/N |
| 5. Is the filler hole unsanitary, or is the lid missing? | Y/N |
| 6. Is the delivery hose nozzle dirty or stored unsafely? | Y/N |
| *Domestic storage tanks* | |
| 7. Can contaminants (e.g. soil on the inside of the lid) enter the tank during filling? | Y/N |
| 8. Does the tank lack a cover? | Y/N |
| 9. Does the tank need a tap for withdrawal of water? | Y/N |
| 10. Is there stagnant water around the storage tank? | Y/N |
| Total score of risks .................... | /10 |

Contamination risk score: 9–10 = very high; 6–8 = high; 3–5 = intermediate; 0–2 = low

**III Results and recommendations**

The following important points of risk were noted: ................................. (list nos 1–10) and the authority advised on remedial action.

Signature of sanitarian ........................................

***Fig. A2.7* Example of sanitary inspection form for deep borehole with mechanical pump**

*Note*: MSD = minimum safe distance determined locally; see section 6.2.2.

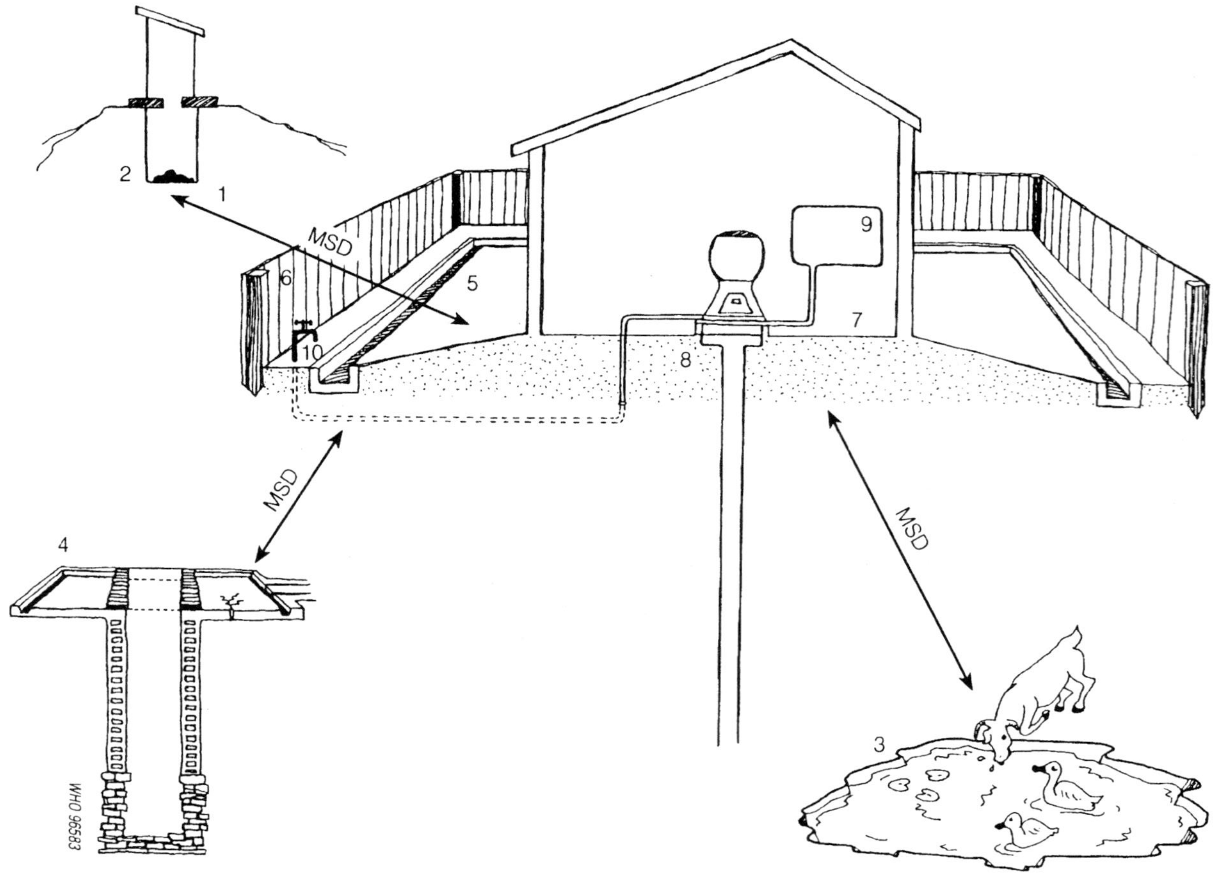

**I Type of facility DEEP BOREHOLE WITH MECHANICAL PUMP**

1. General information: Health centre ..........................................................................
   Village ...................................................................................
2. Code no.—Address ...................................................................................................
3. Water authority/community representative signature .........................
4. Date of visit .......................................
5. Is water sample taken? .... Sample no. ......... Thermotolerant coliform grade .........

| **II Specific diagnostic information for assessment** | **Risk** |
|---|---|
| 1. Is there a latrine or sewer within 15–20 m of the pumphouse? | Y/N |
| 2. Is the nearest latrine a pit latrine that percolates to soil, i.e. unsewered? | Y/N |
| 3. Is there any other source of pollution (e.g. animal excreta, rubbish, surface water) within 10 m of the borehole? | Y/N |
| 4. Is there an uncapped well within 15–20 m of the borehole? | Y/N |
| 5. Is the drainage area around the pumphouse faulty? Is it broken, permitting ponding and/or leakage to ground? | Y/N |
| 6. Is the fencing around the installation damaged in any way which would permit any unauthorized entry or allow animals access? | Y/N |
| 7. Is the floor of the pumphouse permeable to water? | Y/N |
| 8. Is the well seal unsanitary? | Y/N |
| 9. Is the chlorination functioning properly? | Y/N |
| 10. Is chlorine present at the sampling tap? | Y/N |
| Total score of risks .................... | /10 |

Contamination risk score: 9–10 = very high; 6–8 = high; 3–5 = intermediate; 0–2 = low

**III Results and recommendations**

The following important points of risk were noted: ................................. (list nos 1–10) and the authority advised on remedial action.

Signature of sanitarian ........................................

***Fig. A2.8 Example of sanitary inspection form for protected spring source***

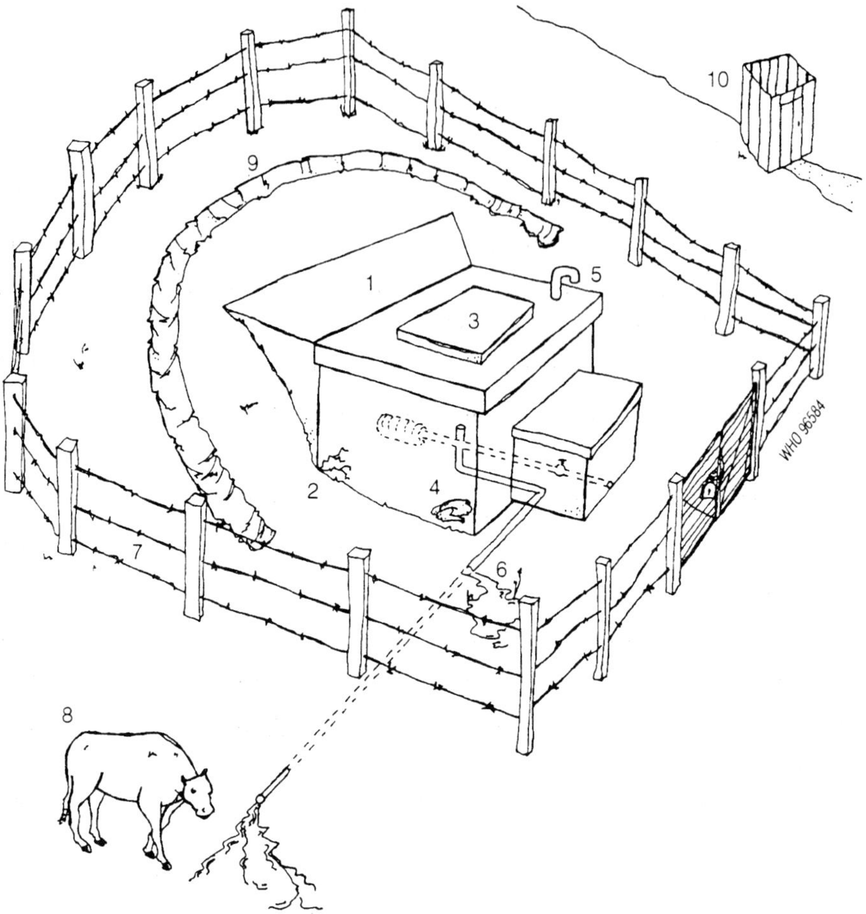

**I Type of facility PROTECTED SPRING SOURCE**

1. General information: Health centre ..........................................................................
   Village ......................................................................................
2. Code no.—Address ..........................................................................................
3. Water authority/community representative signature .........................
4. Date of visit .....................................
5. Water sample taken?....... Sample no......... Thermotolerant coliform grade .........

| **II Specific diagnostic information for assessment** | **Risk** |
|---|---|
| 1. Is the spring source unprotected by masonry or concrete wall or spring box and therefore open to surface contamination? | Y/N |
| 2. Is the masonry protecting the spring source faulty? | Y/N |
| 3. If there is a spring box, is there an unsanitary inspection cover in the masonry? | Y/N |
| 4. Does the spring box contain contaminating silt or animals? | Y/N |
| 5. If there is an air vent in the masonry, is it unsanitary? | Y/N |
| 6. If there is an overflow pipe, is it unsanitary? | Y/N |
| 7. Is the area around the spring unfenced? | Y/N |
| 8. Can animals have access to within 10 m of the spring source? | Y/N |
| 9. Does the spring lack a surface water diversion ditch above it, or (if present) is it nonfunctional? | Y/N |
| 10. Are there any latrines uphill of the spring? | Y/N |

Total score of risks .................... /10

Contamination risk score: 9–10 = very high; 6–8 = high; 3–5 = intermediate; 0–2 = low

**III Results and recommendations**

The following important points of risk were noted: ................................. (list nos 1–10) and the authority advised on remedial action.

Signature of sanitarian ........................................

***Fig. A2.9 Example of sanitary inspection form for surface source and abstraction***

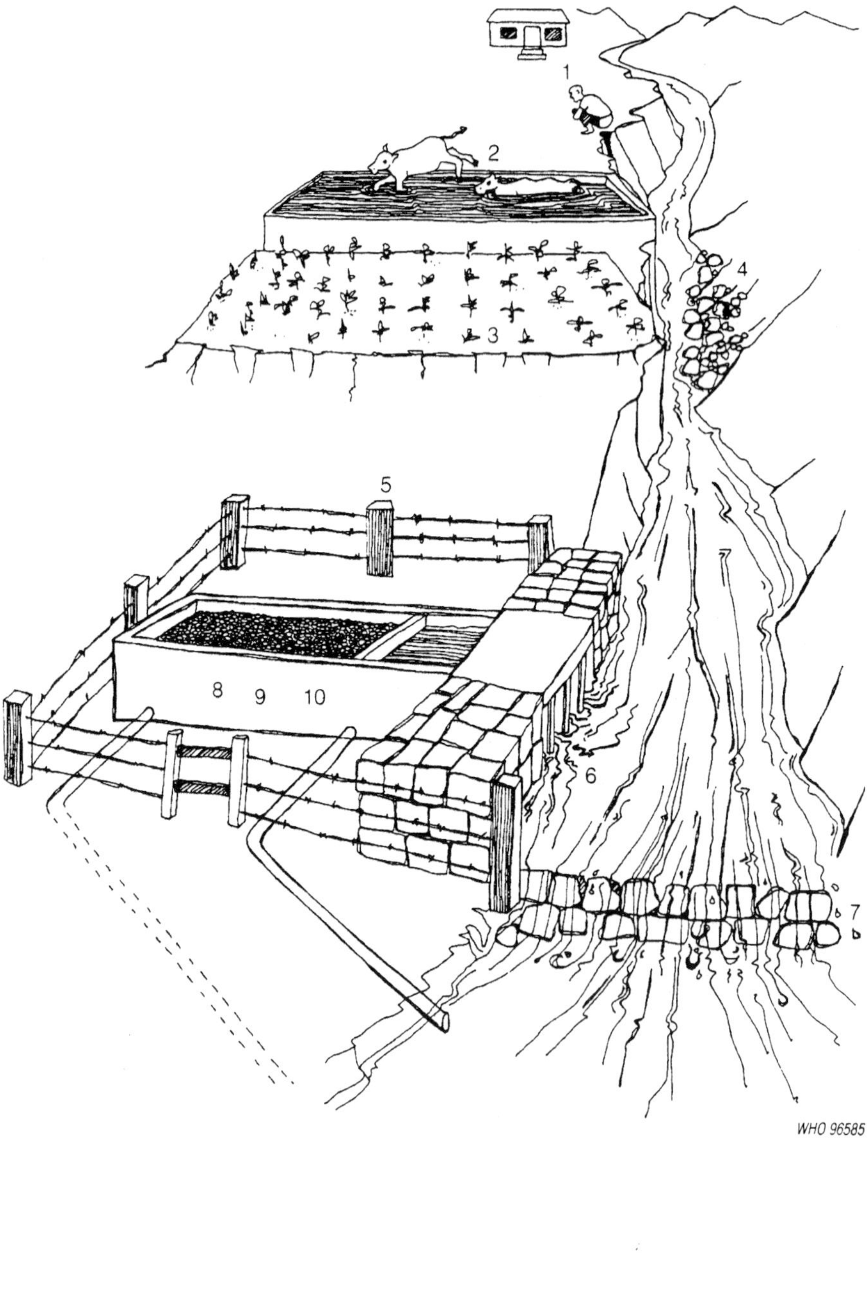

## I Type of facility SURFACE SOURCE AND ABSTRACTION

1. General information: Health centre ........................................................
   Village ........................................................
2. Code no.—Address ........................................................
3. Water authority/community representative signature .........................
4. Date of visit .....................................
5. Water sample taken?....... Sample no......... Thermotolerant coliform grade .........

## II Specific diagnostic information for assessment

| | | Risk |
|---|---|---|
| 1. | Is there any human habitation upstream, polluting the source? | Y/N |
| 2. | Are there any farm animals upstream, polluting the source? | Y/N |
| 3. | Is there any crop production or industrial pollution upstream? | Y/N |
| 4. | Is there a risk of landslide or mudflow (causing deforestation) in the catchment area? | Y/N |
| 5. | Is the intake installation unfenced? | Y/N |
| 6. | Is the intake unscreened? | Y/N |
| 7. | Does the abstraction point lack a minimum-head device (weir or dam to ensure minimum head of water)? | Y/N |
| 8. | Does the system require a sand or gravel filter? | Y/N |
| 9. | If there is a filter, is it functioning badly? | Y/N |
| 10. | Is the flow uncontrolled? | Y/N |

Total score of risks .................... /10

Contamination risk score: 9–10 = very high; 6–8 = high; 3–5 = intermediate; 0–2 = low

## III Results and recommendations

The following important points of risk were noted: ................................. (list nos 1–10) and the authority advised on remedial action.

Signature of sanitarian .......................................

***Fig. A2.10 Example of sanitary inspection form for piped distribution***

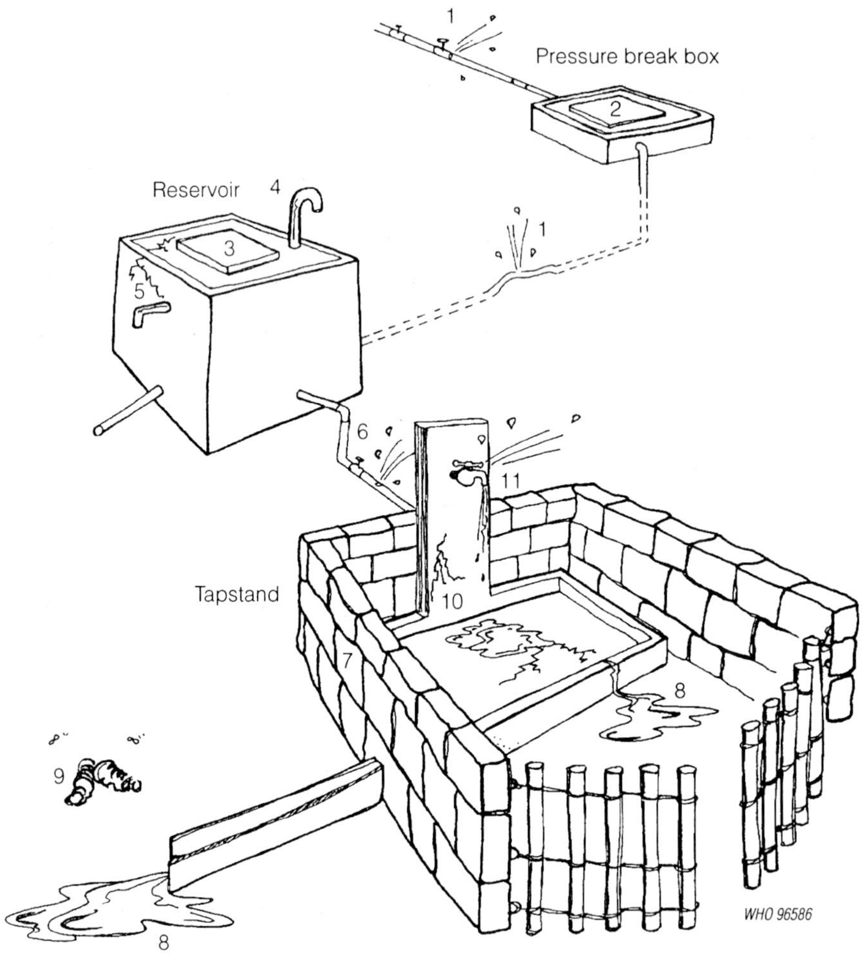

**I Type of facility PIPED DISTRIBUTION**

1. General information: Health centre ......................................................................
   : Village .............................................................................
2. Code no.—Address .............................................................................................
3. Water authority/community representative signature .........................
4. Date of visit .....................................
5. Water sample taken?....... Sample no......... Thermotolerant coliform grade .........

| **II Specific diagnostic information for assessment** | **Risk** |
|---|---|
| 1. Is there any point of leakage between source and reservoir? | Y/N |
| 2. If there are any pressure break boxes, are their covers unsanitary? | Y/N |
| If there is a *reservoir*: | |
| 3. Is the inspection cover unsanitary? | Y/N |
| 4. Are any air vents unsanitary? | Y/N |
| 5. Is the reservoir cracked or leaking? | Y/N |
| 6. Are there any leaks in the distribution system? | Y/N |
| 7. Is the area around the tapstand unfenced (dry stone wall and/or fencing incomplete)? | Y/N |
| 8. Does water accumulate near the tapstand (requires improved drainage canal)? | Y/N |
| 9. Are there human excreta within 10 m of the tapstand? | Y/N |
| 10. Is the plinth cracked or eroded? | Y/N |
| 11. Does the tap leak? | Y/N |

Contamination risk score: 10–11 = very high; 6–9 = high; 3–5 = intermediate; 0–2 = low

**III Results and recommendations**

The following important points of risk were noted: ................................. (list nos 1–11) and the authority advised on remedial action.

Signature of sanitarian .........................................

### Fig. A2.11 Example of sanitary inspection form for water-treatment plant

**I General information** WATER-TREATMENT PLANT

1. Date of survey............/.........../...........
2. Survey of ............... Source ............... Intake ............................ Treatment plant ...................... Distribution
3. Carried out by ................................ Name of person ................................ Agency
4. Name of supply ............................... State ............................ District ............................... Treatment plant
5. Address ..............................................................................................................
6. Person in charge ..................................................................................................
7. Year started operation ..........................................................................................
8. Area served .................................. Population served ...........................................
9. Treatment-plant capacity Designed ............... Actual ................................
10. Security of plant Fence: Y/N Security guard: Y/N

**II Source**

1. Type of water source: ............... Reservoir ............. Stream ........... River ........... Well ................ Others

**III Intake**

1. Is the intake adequate with respect to:

| | |
|---|---|
| Location? | Y/N |
| Structure? | Y/N |
| Maintenance? | Y/N |
| Pollution sources in the vicinity? | Y/N |

**IV Treatment processes employed**

1. Fine screen ........................
2. Grit chamber ........................
3. Oil and grease trap ........................
4. Presedimentation ........................
5. Predisinfection/oxidation ........................ Chlorine ........................ Ozone
6. Activated carbon treatment ........................

| | | | | |
|---|---|---|---|---|
| 7. | Aeration | ........................ | | |
| 8. | Coagulation and flocculation | ................. Lime | ................. Alum | ................. Others |
| 9. | Sedimentation | ........................ Rectangular | ........................ Circular | ........................ Others |
| 10. | Filtration | ........................ Slow | ........................ Rapid | ........................ Granular carbon |
| 11. | Disinfection | ........................ Chlorine | ........................ Ozone | ........................ Others |

12. Other processes (specify): ..........................................................................................

..........................................................................................................................

## V Sedimentation

1. No. of sedimentation tank: ..........................................................................
2. Frequency of desludging: .............................................................................
3. Type of desludging facility: ..........................................................................
4. Method of sludge disposal: ..........................................................................
5. General appearance of clarified water: ..........................................................
6. Turbidity (NTU) at inlet: .......................... (NTU) at outlet: ...............................

## VI Filtration

1. No. of filters: ................................................................................................
2. Filtration rate: ..............................................................................................
3. Filter run: .....................................................................................................
4. Depth of gravel: ...........................................................................................
5. Depth of sand: .............................................................................................

## VII Backwashing

1. Criteria used for initiating backwashing:

| | | |
|---|---|---|
| Air scour: | ........................ Rate | ........................ Duration |
| Water scour: | ........................ Rate | ........................ Duration |

2. Distribution of air and water supply in the sand bed:

   .................... ........................
   Even Uneven

3. Capacity of clean water for backwash: ..................................................................

4. Any mud balls or cracks in the filter bed?

   Before backwash .......................

   After backwash .......................

5. Where does the wash water go? ...........................................................................

**VIII Fluoridation**

1. Chemical used: ...........................................................................................

2. Dosage of chemical: .....................................................................................

**IX Chlorination**

1. Any interruption in chlorination? ..........................................................................

2. Frequency of interruption: ..................................................................................

3. Cause of interruption: ......................................................................................

4. Type of chemical used: ......................................................................................

5. Dosage of chemical: ........................................................................................

6. Safety equipment and measures: .............................................................................

7. Reserve stock of disinfectant: ............................................... Quantity ................

8. Storage conditions: .........................................................................................

**X Clear-water tank(s)**

1. No. of tanks: ...............................................................................................

2. Capacity of each tank: ......................................................................................

3. Concentration of free residual chlorine: ....................................................................

4. pH: .........................................................................................................

5. Chemical used for pH adjustment and its dosage: ............................................................

6. Any leak in the tank? .......................................................................................

7. Is the tank properly covered and locked? ....................................................................

8. Any scum or foreign substances in the tank? ................................................................

9. Are air vents and overflow pipes protected by screens? ......................................

## XI Process control

| | Yes | No | Frequently |
|---|---|---|---|
| 1. Jar test: | .................. | .................. | .................. |
| 2. pH: | .................. | .................. | .................. |
| 3. Free residual chlorine: | .................. | .................. | .................. |
| 4. Colour: | .................. | .................. | .................. |
| 5. Turbidity: | .................. | .................. | .................. |
| 6. *E.coli*/thermotolerant coli: | .................. | .................. | .................. |
| 7. Fluoride: | .................. | .................. | .................. |
| 8. Others: | .................. | .................. | .................. |

SANITARY INSPECTION

## XII Record keeping

1. Chemical consumption: ..........................................................................
2. Process-control tests: ..........................................................................
3. Bacteriological examination: ..........................................................................
4. Residual chlorine: ..........................................................................
5. Others: ..........................................................................

## XIII Maintenance

| | Cleaning | Calibrating/oiling/ greasing |
|---|---|---|
| 1. Screen: | ...................................... | ...................................... |
| 2. Pumping facility: | ...................................... | ...................................... |
| 3. Chlorine-dosing facility: | ...................................... | ...................................... |
| 4. Alum-dosing facility: | ...................................... | ...................................... |
| 5. Fluoride-dosing facility: | ...................................... | ...................................... |
| 6. Instrument (gauge, recording devices, etc.): | ...................................... | ...................................... |
| 7. General housekeeping: | ...................................... | ...................................... |
| 8. Storage of chemicals: | ...................................... | ...................................... |
| | Adequate | Inadequate |

**XIV Personnel**

1. No. of present staff: .......................... ..........................
   Permanent Casual

2. Academic level of the plant superintendent or the most senior operator of the treatment plant:

   ..............................................................................................................................

3. Length of service in present water-treatment plant: ..............................................

4. Total experience in water treatment: .....................................................................

**XV Complaints received**

1. From operators: ...........................................................................................

   ..............................................................................................................................

   ..............................................................................................................................

   ..............................................................................................................................

   ..............................................................................................................................

2. From management: ......................................................................................

   ..............................................................................................................................

   ..............................................................................................................................

   ..............................................................................................................................

**XVI Problems (if any) with:**

| | Yes | No | Description of problems |
|---|---|---|---|
| 1. Fine screen: | ................ | ................ | .......................................... |
| 2. Grit chamber: | ................ | ................ | .......................................... |
| 3. Oil and grease trap: | ................ | ................ | .......................................... |
| 4. Presedimentation: | ................ | ................ | .......................................... |
| 5. Activated carbon: | ................ | ................ | .......................................... |
| 6. Aeration: | ................ | ................ | .......................................... |
| 7. Coagulation and flocculation: | ................ | ................ | .......................................... |
| 8. Sedimentation: | ................ | ................ | .......................................... |
| 9. Filtration: | ................ | ................ | .......................................... |
| 10. Fluoridation: | ................ | ................ | .......................................... |

| | | | |
|---|---|---|---|
| 11. Disinfection: | ................ | ................ | .......................................... |
| 12. Other process: | ................ | ................ | .......................................... |
| 13. Process control: | ................ | ................ | .......................................... |
| 14. Record keeping: | ................ | ................ | .......................................... |
| 15. Maintenance: | ................ | ................ | .......................................... |

**XVII Flow diagram of water works** (*insert diagram*)

**XVIII Remedial measures recommended**

1. Measures to be taken immediately:

.............................................................................................................................

.............................................................................................................................

.............................................................................................................................

.............................................................................................................................

.............................................................................................................................

.............................................................................................................................

2. Measures to be taken later on:

.............................................................................................................................

.............................................................................................................................

.............................................................................................................................

.............................................................................................................................

.............................................................................................................................

**XIX Have problems identified in the previous sanitary survey been corrected?**

.............................................................................................................................

.............................................................................................................................

.............................................................................................................................

.............................................................................................................................

.............................................................................................................................

.............................................................................................................................

Signature of inspector: ...............................................................................................

Annex 3

# Examples of possible responsibilities of surveillance staff

## A3.1 National surveillance team

Members of the national surveillance team may be responsible for:

- advising at the highest level on surveillance policy and strategy to ensure the maintenance and development of suitable supplies of safe water;
- formulating and revising technical standards for the control of drinking-water quality;
- coordinating the supervision, control, and evaluation of local surveillance staff and operator-level quality-control staff, where appropriate;
- coordinating and promoting the development of water surveillance at all levels;
- promoting and advising on the establishment of laboratories;
- supporting and coordinating the training of staff;
- developing and managing a national database for strategic planning purposes;
- attending annual meetings with the national planning authority and appropriate water-supply and construction authorities to discuss and agree sector plans.

## A3.2 Provincial water surveillance coordinator

Responsibilities of the provincial water surveillance coordinator may include:

- planning and coordinating annual water-surveillance programmes with district coordinators and the provincial head of environmental health;
- coordinating the supply of equipment and consumables;
- making regular (e.g. quarterly) supervisory visits to each district, accompanying the district coordinator on follow-up visits to make spot checks on sanitarians' performance of duties, and noting deficiencies on visit reports;
- detecting errors in reporting and methodology and correcting them;
- collecting and checking monthly surveillance reports from districts and maintaining an up-to-date surveillance database;
- informing the provincial head about priorities for improvement, results not meeting water-quality standards, and progress with surveillance;
- sending summary surveillance reports to the provincial water-supply agencies; where necessary, drawing to their attention quality and service levels

presenting a risk to the health of the consumer; deciding on the remedial action to be taken with the appropriate authorities;
- coordinating periodic meetings with senior staff of water-supply agencies to discuss the scope of, and dividing lines between, quality-control and surveillance duties;
- formulating advice for emergencies and proposing medium-term provincial strategies to remedy deficiencies in water-supply services that will reduce the risk to the consumer;
- preparing annual reports on all surveillance, monitoring, and quality-control activities in districts; identifying in these reports the areas at greatest risk, and the deficiencies in the numbers, competence, and training of surveillance staff;
- coordinating programme-evaluation and retraining meetings;
- promoting remedial action and good operation and maintenance strategies;
- assessing the workload of the district and provincial water laboratories, and coordinating analytical quality control and the referral of samples both between these laboratories themselves and between them and those at national level;
- arranging for monthly reports to be sent to the national surveillance team.

## A3.3 District water surveillance coordinators

District water surveillance coordinators may be responsible for:
- planning and coordinating the annual water-surveillance programme with the district surveillance team, the head of environmental health, and the provincial surveillance coordinator;
- supervising and making spot-checks on sanitarians' surveys by making field visits to urban and rural areas; reporting findings to the provincial surveillance coordinator and then to the head of environmental health;
- validating reports and quality results, deciding whether or not emergencies exist, and verifying "odd" results by making follow-up visits to the field;
- collecting monthly urban and rural surveillance reports; maintaining and updating an archive of data including an inventory of all water supplies, piped coverage levels, and rural piped and unpiped supplies, and a similar archive covering sanitation (if this is also the responsibility of the surveillance agency);
- obtaining water quality reports from hospital laboratories and ensuring that the results are compared with the appropriate sanitary inspection report;
- discussing with the provincial water surveillance coordinator (and ultimately the head of environmental health) both routine and anomalous results, and identifying and reporting high-risk communities;
- sending urban water-surveillance reports to managers of the local water authorities;

- meeting urban water-supply operators and managers, identifying high-risk zones in their supply, and agreeing joint quality-control arrangements; drawing the managers' attention to risks and suggesting emergency action, where appropriate;
- giving advice on emergency measures, including warnings to the public, and agreeing responsibilities for action with other concerned agencies;
- arranging for routine monthly reports to be sent to the provincial water surveillance coordinator;
- coordinating and supporting community-based hygiene education activities and training in sanitary inspection for community-based volunteers;
- keeping records of community volunteers and encouraging community involvement in water-supply surveillance and improvement;
- coordination of training for sanitary technicians in work with, and provision of advice to, communities;
- checking that sanitary technicians are providing good technical advice and support to the community for remedial action and improvement;
- making spot-checks to ensure that recommendations for remedial action are acted on, and reporting deficiencies to the provincial surveillance coordinator and thence to the head of environmental health;
- participating in annual intersectoral strategic planning meetings for improving water-supply services; presenting evidence of the need for improvement in specific areas;
- investigating water-related outbreaks of disease and arranging for emergency action for community protection;
- providing an annual report on urban and rural levels of sanitation service.

## A3.4 Water surveillance sanitary technicians

Water surveillance sanitary technicians may be responsible for:

- carrying out routine (e.g. weekly) monitoring of water-distribution systems, including fixed-point and random sampling;
- checking and recording chlorine residuals on the spot, and sampling from sites showing low levels (e.g. <0.1 mg/litre free chlorine) for bacteriological analysis; transporting samples to the appropriate laboratory;
- entering analytical results in surveillance reports and making weekly reports to the surveillance coordinator;
- intensifying the monitoring of high-risk water-supply zones, such as those where pressure is low, leakage high, the results of bacteriological tests bad, or standpipes are used;
- carrying out special sampling programmes in periurban and urban areas unserved by piped systems and preparing reports on them;
- informing the surveillance coordinator and head of environmental health of

high-risk zones as soon as they are identified, and indicating by appropriate means any advice to be given to the community in an emergency;
- periodically providing samples to the provincial laboratory for chemical analysis and obtaining the results for inclusion in the district archive;
- liaising with local treatment-plant operators and making spot-checks to ensure that they are keeping adequate daily records; noting deficiencies and entering them on surveillance reports;
- maintaining a register of all major sources of pollution of water resources, and carrying out periodic surveys of these water resources (where this is the responsibility of the surveillance agency);
- taking samples of water from urban water sources, and sending them to the appropriate laboratory for full analysis;
- undertaking water source surveys;
- carrying out sanitary surveys of community water supplies;
- providing summary advisory reports to community representatives, pointing out essential remedial action and, wherever possible, providing technical support for improvement;
- keeping and extending an inventory of all water sources and their location, together with a sanitation inventory (where this is the responsibility of the surveillance agency);
- preparing a monthly summary of all sanitary surveys, including the advice provided on remedial action, and sending this summary to the district surveillance coordinator;
- notifying the district-level surveillance coordinator of high-risk facilities, and requesting support from the coordinator for follow-up inspection and analysis;
- drawing up an annual programme of hygiene education, and requesting the coordinator to provide the necessary materials and technical support for its implementation;
- developing and implementing a training programme for community-level surveillance of water resources and source protection, and requesting the coordinator to provide the necessary technical support and materials;
- liaising with community surveillance volunteers, receiving their reports, and providing advice and training.

Annex 4

# Sampling methods for bacteriological testing

When water samples are collected for analysis, care should be taken to ensure that there is no external contamination of the samples. Unless valid samples are collected, the results of the subsequent analysis may be misleading.

Several types of bottle may be used for sampling, but glass bottles are best. These should have securely fitting stoppers or caps with nontoxic liners, and both bottles and stoppers should be sterilized. Each cap should have a metal sleeve clear of the screw thread to ensure that the risk of contaminating the water sample is minimized. Cotton wool plugs and paper caps should be avoided as they tend to fall off during and after sampling and increase the risk of contamination. The bottles should hold at least 200 ml of water.

Whenever chlorine is used for disinfection, a chlorine residual may be present in the water after sampling and will continue to act on any bacteria in the sample; the results of the microbiological analysis may therefore not be indicative of the true bacteriological content of the water. To overcome this difficulty, it is common procedure to add sodium thiosulfate to the sample, which immediately inactivates any residual chlorine but does not affect the microorganisms that may be present. The sodium thiosulfate should be added to the sample bottles before they are sterilized. For 200-ml samples, four or five drops of aqueous sodium thiosulfate solution (100 g/litre) should be added to each clean sample bottle. The stopper is loosely inserted into the bottle, and a brown paper or aluminium foil cover is tied to the neck of the bottle to prevent dust from entering. The bottle is then sterilized in a hot-air oven for 1 hour at 160 or 170 °C for 40 minutes or in an autoclave at 121 °C for 20 minutes. If no other facilities are available, a portable sterilizer or pressure cooker can be used, but sterilization will then take 30–45 minutes. To prevent the stopper from getting stuck during sterilization, a strip of brown paper (75 × 10 mm) should be inserted between the stopper and the neck of the bottle.

For reasons of cost, bottles should be reused. After the samples have been analysed in the regional or central laboratory, bottles should be resterilized and, if possible, returned to the sender.

Water can be divided into three basic types for the purpose of sampling:

- water from a tap in a distribution system or from a fixed pump outlet, etc.
- water from a watercourse (river, lake, etc.) or a tank
- water from a dug well, etc., where sampling is more difficult than from an open watercourse.

## A4.1 Sampling from a tap or pump outlet

**A. Clean the tap**
Remove from the tap any attachments that may cause splashing. Using a clean cloth, wipe the outlet to remove any dirt.

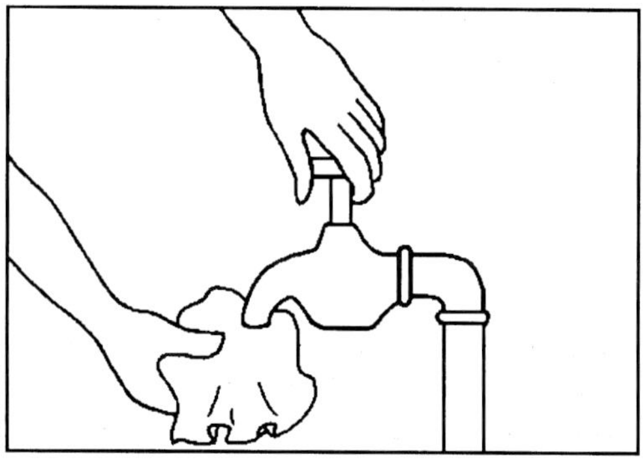

WHO 96587

**B. Open the tap**
Turn on the tap at maximum flow and let the water run for 1–2 minutes.
*Note*: Some investigators do not continue to stages C and D but take the sample at this stage; in this case, the tap should not be adjusted or turned off, but left to run at maximum flow. The results obtained in this way will provide information on the quality of the water as consumed. If the procedure is continued to stages C and D, however, the results represent the quality of the water excluding contamination by the tap.

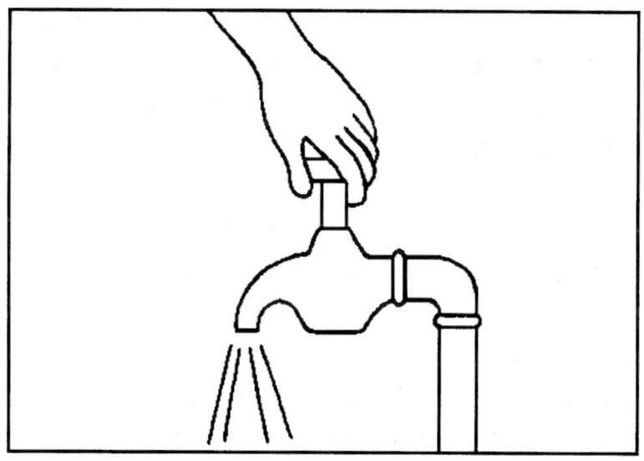

WHO 96588

**C. Sterilize the tap**
Sterilize the tap for a minute with the flame from a gas burner, cigarette lighter, or an ignited alcohol-soaked cotton-wool swab.

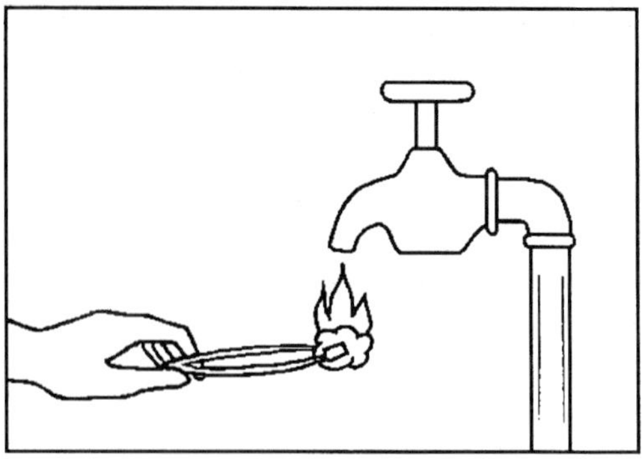

WHO 96589

**D. Open the tap before sampling**
Carefully turn on the tap and allow the water to flow for 1–2 minutes at a medium flow rate. Do not adjust the flow after it has been set.

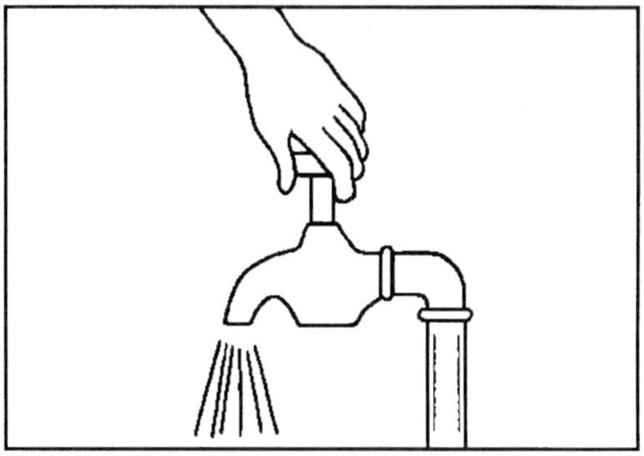

WHO 96590

**E. Open the sterilized bottle**
Take out a bottle and carefully unscrew the cap or pull out the stopper.

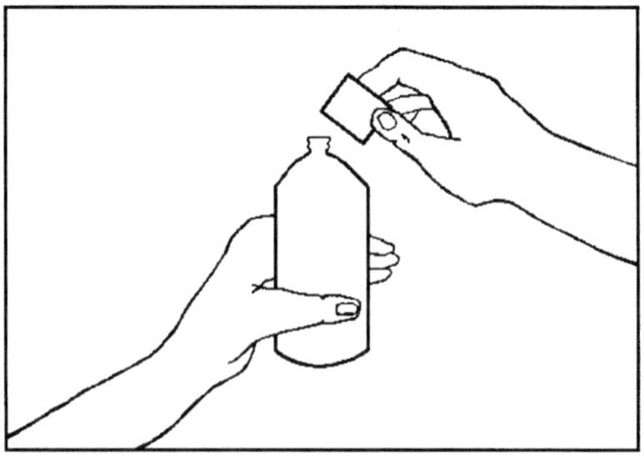

WHO 96591

**F. Fill the bottle**
While holding the cap and protective cover face downwards (to prevent entry of dust, which may contaminate the sample), immediately hold the bottle under the water jet, and fill.

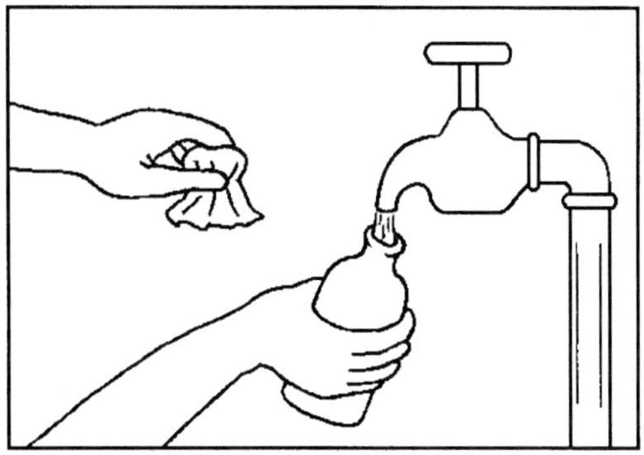
WHO 96592

A small air space should be left to make shaking before analysis easier.

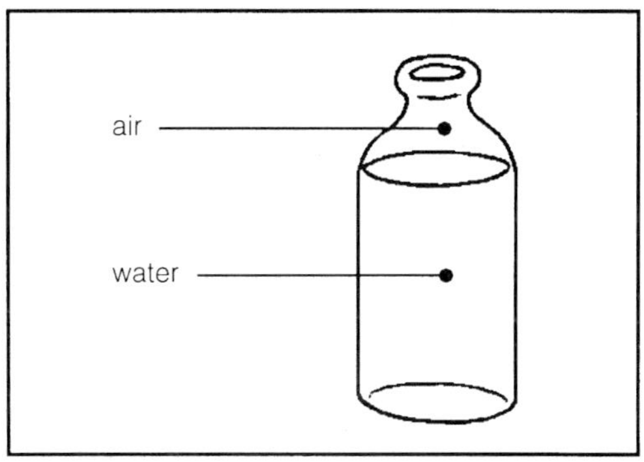

WHO 96593

**G. Stopper or cap the bottle**
Place the stopper in the bottle or screw on the cap and fix the brown paper protective cover in place with the string.

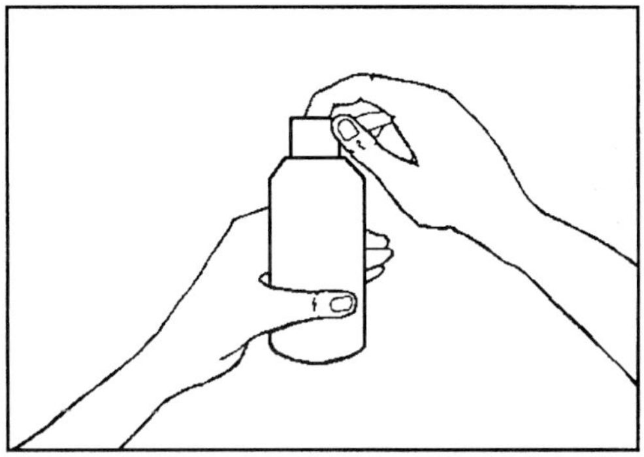
WHO 96594

## A4.2 Sampling from a watercourse or reservoir

Open the sterilized bottle as described in section A4.1.

**A. Fill the bottle**

Holding the bottle by the lower part, submerge it to a depth of about 20 cm, with the mouth facing slightly upwards. If there is a current, the bottle mouth should face towards the current.

The bottle should then be capped or stoppered as described previously.

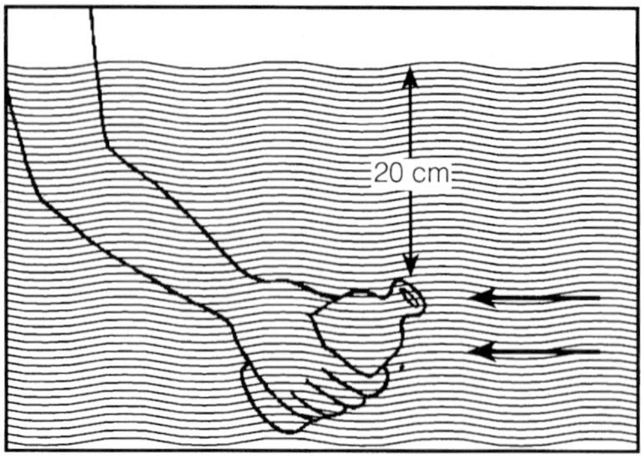


WHO 96595

## A4.3 Sampling from dug wells and similar sources

**A. Prepare the bottle**

With a piece of string, attach a clean weight to the sampling bottle.

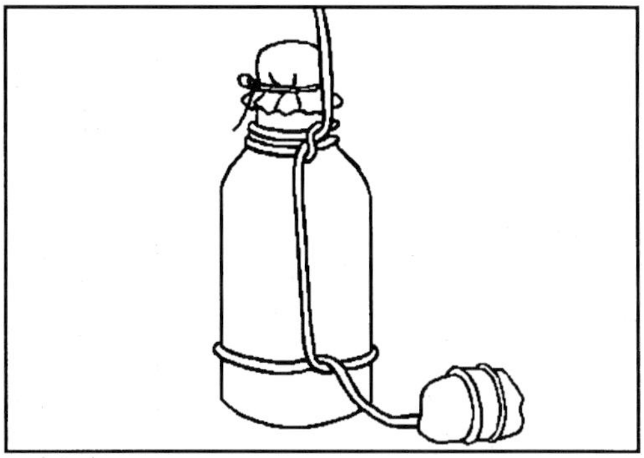

WHO 96596

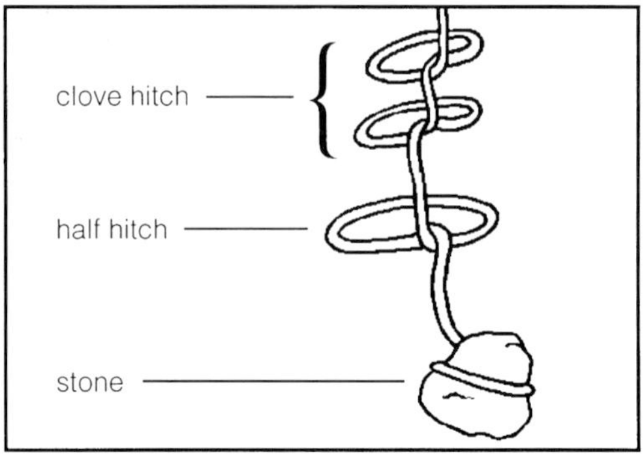


WHO 96597

**B. Attach the bottle to the string**
Take a 20-m length of clean string rolled around a stick and tie it to the bottle string. Open the bottle as described in section A4.1.

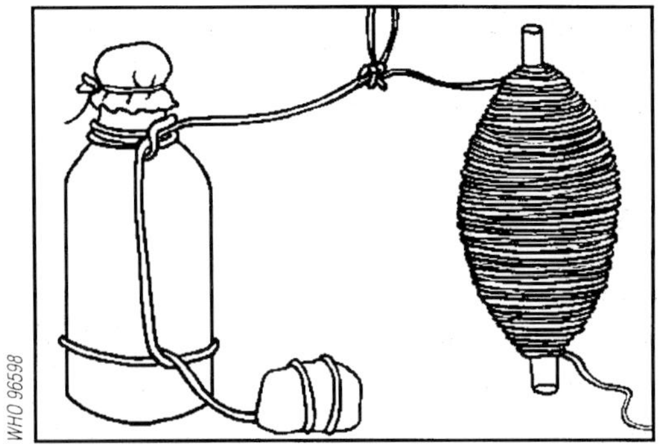


**C. Lower the bottle**
Lower the bottle, weighed down by the weight, into the well, unwinding the string slowly. Do not allow the bottle to touch the sides of the well.

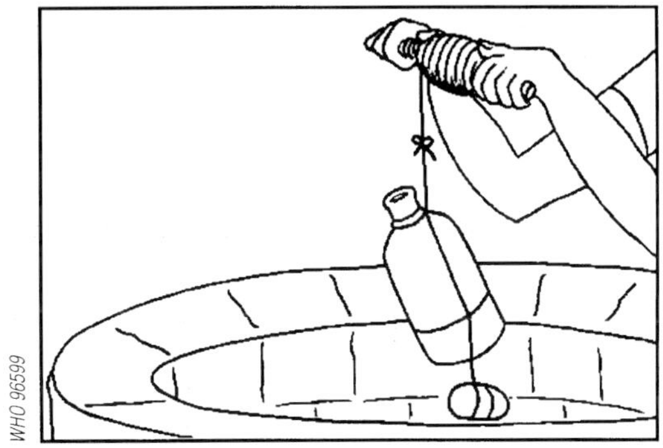


**D. Fill the bottle**
Immerse the bottle completely in the water and lower it well below the surface without hitting the bottom or disturbing any sediment.

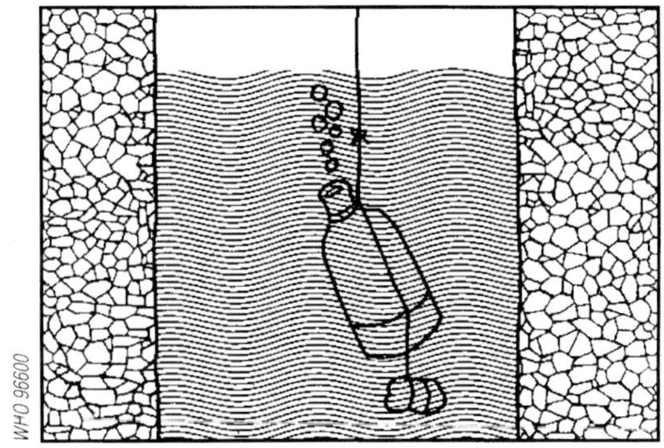

**E. Raise the bottle**

Once the bottle is judged to be filled, rewind the string on the stick to bring up the bottle. If the bottle is completely full, discard some water to provide an air space. Stopper or cap the bottle as described previously.

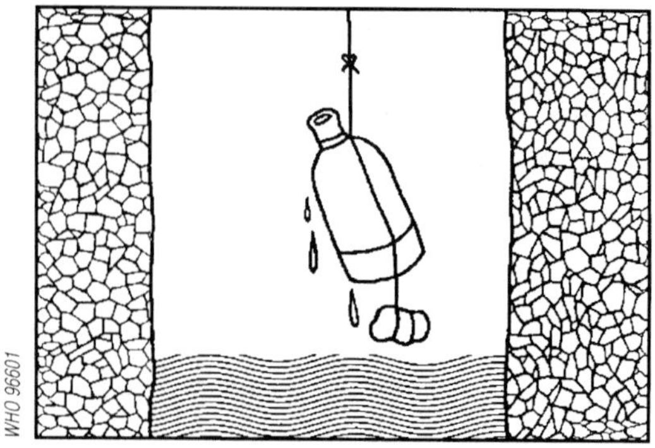

Annex 5

# Multiple-tube method for thermotolerant (faecal) coliforms

In the multiple-tube method, a series of tubes containing a suitable selective broth culture medium is inoculated with test portions of a water sample. After a specified incubation time at a given temperature, each tube showing gas formation is regarded as "presumptive positive" since the gas indicates the possible presence of coliforms. However, gas may also be produced by other organisms, and so a subsequent confirmatory test is essential. The two tests are known respectively as the *presumptive test* and the *confirmatory test*.

For the confirmatory test, a more selective culture medium is inoculated with material taken from the positive tubes. After an appropriate incubation time, the tubes are examined for gas formation as before. The most probable number (MPN) of bacteria present can then be estimated from the number of tubes inoculated and the number of positive tubes obtained in the confirmatory test, using specially devised statistical tables. This technique is known as the MPN method.

## A5.1 Inoculation

Different test portions to provide tenfold serial dilution steps may be used, the dilutions being based on the anticipated number of coliform bacteria in the water sample being tested. The reliability of the result obtained depends on the number of tubes inoculated with each test portion; in certain instances, the number can be reduced to three in each dilution step. Each combination of inoculated tubes will have its own table of MPN values. Typical volumes for analysis are given in Table 4.4 (p. 62).

## A5.2 Unpolluted and treated water

Water in or entering the distribution system may generally be assumed to contain little or no pollution. In this case, it is recommended that one 50-ml plus five 10-ml volumes of water sample should be inoculated into the tubes; five tubes should each contain 10 ml and one tube 50 ml of double-strength medium.

## A5.3 Polluted water

Water suspected to be more highly contaminated, e.g. untreated water from certain raw water sources, should be examined using different inoculation volumes in tenfold dilution steps. The following inoculations are normally made:

— 10 ml of sample to each of five tubes containing 10 ml of double-strength medium;
— 1.0 ml of sample to each of five tubes containing 10 ml of single-strength medium;
— 1.0 ml of a 1 : 10 dilution of sample (i.e. 0.1 ml of sample) to each of five tubes containing 10 ml of single-strength medium.

If the sample is expected to be highly contaminated, aliquots of 1.0 ml of tenfold serial dilutions from each dilution step are inoculated into five tubes that each contain 10 ml of single-strength medium.

If the workload is very heavy and the time available is limited, the number of tubes can be reduced to three in each series. Statistically, however, inoculation of five tubes with each sample volume produces a more reliable MPN result.

## A5.4 Equipment and supplies

The following laboratory equipment and glassware are essential:

- *Autoclave*: required for sterilizing the culture media. Its size should be determined by the number and type of samples to be taken. Operation of the autoclave should be strictly in accordance with the manufacturer's instructions and should ensure that all the air in the chamber is replaced by steam. Sterilization should be achieved in not more than 30 minutes. Strict adherence to recommended sterilization temperatures and times for different types of culture media is essential. Racks are needed to hold tubes and bottles of prepared culture media in the autoclave.
- *Incubator(s) or water-baths*: must each be fitted with a temperature control and should be capable of maintaining a uniform temperature correct to 35 or 37 ± 0.5 °C and/or 44 or 44.5 ± 0.25 °C. The choice of temperature depends on the indicator bacteria and the medium used. The temperature of incubators and water-baths fitted with thermometers placed at representative points should be monitored periodically (preferably daily). Stainless-steel racks should be fitted to hold sample tubes.
- *Balance*: needed for weighing powdered culture medium. It should have an accuracy of 0.05 g. A weighing scoop is also required. (No balance is required if culture media are available in suitable preweighed quantities.)
- *Water distillation apparatus, hose, and container*: required to produce non-toxic water, i.e. water free from any substances that can interfere with bacterial growth. The container for the distilled water should have a volume of at least 5 litres and be fitted with a tap.
- *Pipettes*: 1 ml and 10 ml, with cotton plugs at the mouthpiece, are required. The 1-ml pipettes should be graduated in 0.1-ml increments and are used for preparing dilutions; the 10-ml pipettes are used for the addition of samples to tubes containing media. Any pipettes with chipped or broken tips should

be discarded. Glass pipettes can be conveniently stored in a sterilizable metal container; alternatively, disposable plastic pipettes can be used. A separate container should be employed for each size of pipette. Pipettes may also be wrapped individually in paper and heat-sterilized. Pipette canisters and bulbs are also needed, as is a container for discarded pipettes.

- *Test-tubes and racks*: tubes can be 20 × 150 mm in size for 10-ml sample volumes plus 10 ml of culture medium (screw caps are not recommended for fermentation media). The racks should be large enough to accommodate culture tubes of the largest diameter used.
- *Bottles*: used for the larger volumes consisting of 50 ml of sample and 50 ml of culture medium. They should have loose-fitting caps and ideally be calibrated with 50-ml and 100-ml marks.
- *Media preparation equipment*: glass or stainless-steel containers (usually flasks) are required. Any heating equipment and stirrers used in the preparation of media should be clean and free from soluble toxic materials.
- *Gas burner*: a Bunsen or similar burner is adequate.
- *Culture tubes containing inverted vials (Durham tubes)*: each tube should be large enough for a vial, completely filled with medium, to be submerged in it.
- *Inoculation loop and holder*: lengths of 24- or 26-gauge wire (7.5–10 cm) should be used. Nichrome wire is acceptable, but platinum–iridium is better. The wire is set in a handle made of metal or glass, of diameter similar to that of a pencil. To make the inoculation loop, the wire is bent to form a circle 3–4 mm in diameter.
- *Dispenser*: for sodium thiosulfate solution (see below).
- *Cleaning and maintenance equipment*: items such as brushes for cleaning tubes, bottles, etc., a waste bin, and a tool kit are required.
- *Safety equipment*: there should be an adequate first-aid kit and a fire extinguisher or other means of fire control in every laboratory.
- *General laboratory equipment*: various sizes of round and Erlenmeyer flask, beakers, stands, glass or unbreakable plastic measuring flasks, spatulas, etc. are required.

The following equipment is also desirable in a laboratory:

- *Refrigerator*: for the storage of prepared culture media.
- *Hot air sterilizer*: for sterilizing pipettes.

The following consumable items are required:

- *Culture medium*: Table A5.1 describes the uses for the various media; see also section A5.5.
- *Laboratory disinfectant*: for cleaning laboratory surfaces and the pipette discard bin.
- *Detergent*: for washing glassware, etc.
- *Sodium thiosulfate solution*: required when chlorinated supplies are tested.

Sodium thiosulfate neutralizes any residual chlorine in samples at the time of collection, preventing it from acting on any microorganisms present in water samples.

- *Autoclave tape.*
- *Diluent*: typical diluents include Ringer's solution and phosphate-buffered saline.

### Table A5.1 Culture media for MPN[a]

| Medium | Uses | Incubator temperature | Remarks |
|---|---|---|---|
| MacConkey broth | Presumptive isolation of coliform bacteria | 35 ± 0.5 °C or 37 ± 0.5 °C | Traditional medium for the presumptive isolation of coliform bacteria by MPN. The quality of bile salts can vary and may affect the result |
| Lauryl tryptose (lactose) broth | Presumptive isolation of coliform bacteria | 35 ± 0.5 °C or 37 ± 0.5 °C | — |
| | Confirmation of thermotolerant coliform bacteria | 44 °C | — |
| Improved formate lactose glutamate medium | Presumptive isolation of coliform bacteria | 35 ± 0.5 °C or 37 ± 0.5 °C | This is a selective medium because it contains chemically defined nutrients that can be utilized only by a limited number of bacterial species. The composition of the medium is complex and special care is required during preparation |
| Brilliant green lactose (bile) broth; EC | Confirmation of coliform bacteria | 35 ± 0.5 °C or 37 ± 0.5 °C | Media for gas production |
| | Confirmation of thermotolerant coliform bacteria | 44 °C | |
| Tryptone water | Production of indole for confirmation of *Escherichia coli* | 44 °C | The formation of indole, detected by the addition of Kovacs reagent[b] to tryptone water after incubation, with gas production from lactose at 44 °C indicates the presence of *E. coli* |

[a] Adapted from ISO 9308-2: 1990. Detection and enumeration of coliform organisms, thermotolerant coliform organisms, and presumptive *Escherichia coli*—Part 2: Multiple tube (most probable number) method.

[b] To make Kovacs reagent, dissolve 5 g *p*-dimethylaminobenzaldehyde in 75 ml amyl (or isoamyl) alcohol, and add 25 ml concentrated hydrochloric acid slowly. Store at 4 °C in the dark.

## A5.5 Culture media and dilution water

Commercially available dehydrated media simplify the preparation of culture broths and are therefore recommended for laboratory work. Various manufacturers produce these media as powders, which can then be easily weighed out, dissolved in distilled water, and dispensed into culture tubes before sterilization.

### A5.5.1 Preparation of media

Media should be prepared in accordance with the manufacturer's instructions, as follows:

(a) Dissolve the stated amount of the dehydrated medium in distilled water to obtain the double-strength or single-strength presumptive medium (for confirmatory analysis, only single-strength medium is used).

(b) Dispense the requisite volume into culture tubes containing an inverted Durham tube, and cap the culture tubes.

(c) Sterilize in an autoclave or pressure cooker at 115 °C for 10 minutes (or in accordance with the manufacturer's specifications). It is particularly important that media containing disaccharides, e.g. lactose, are not autoclaved at higher temperatures.

(d) The sterilized medium may be stored at room temperature (approximately 25 °C) or, ideally, at 2–8 °C. Media should in any case be warmed to room temperature before use to ensure that all components have redissolved. In addition, since several dyes are light-sensitive, the solution should be protected from exposure to light.

### A5.5.2 Preparation of dilution water

A special buffered, sterilized water is used to make sample dilutions for inoculation into the culture medium. It is prepared from a concentrated stock solution of phosphate buffer. To make the stock solution, dissolve 34 g of potassium dihydrogen phosphate ($KH_2PO_4$) in 500 ml of distilled water. The pH should be checked with a pH-meter; it should be 7.2. It can be increased if necessary by adding a few drops of sodium hydroxide solution (4.0 g dissolved in 1000 ml of distilled water). Then add sufficient distilled water to make up to 1 litre. When the stock solution is not in use, it should be stored in a tightly closed bottle at 4–10 °C, to prevent microbial growth.

When using the dilution water, add 1.25 ml of stock phosphate solution to 1 litre of distilled water and dispense into bottles for sterilization in the autoclave. Before sterilization, loosen the stoppers of the bottles. Sterilize for 15 minutes at 121 °C. Tighten the stoppers after sterilization and store the dilution water in a clean place until needed.

An alternative dilution water can be prepared by the addition of magnesium chloride and has been shown to give a slightly higher recovery rate. Other alternatives include a 0.1% solution of peptone in distilled water (final pH 6.8), Ringer's solution, and physiological saline (9g of sodium chloride per litre). These should be sterilized after dispensing into bottles, as described above.

## A5.6 Application to unpolluted water

### A5.6.1 Procedure

The procedure to be used for testing relatively unpolluted water, such as treated water from waterworks, is described below.

**A.** Remove the cap from the sample bottle.

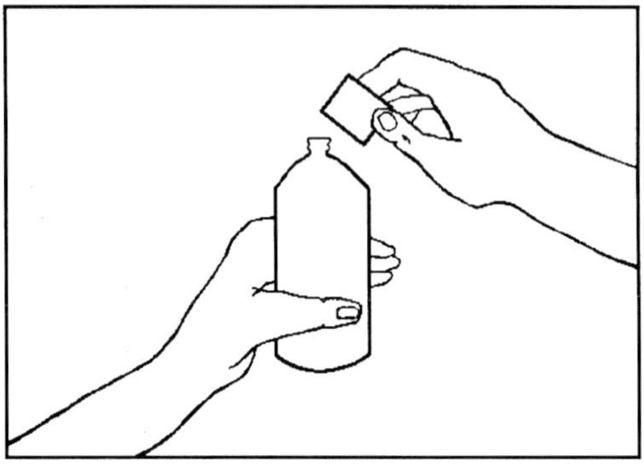

WHO 96602

**B.** With the stopper in position, shake the bottle vigorously to achieve a homogeneous dispersion of bacteria. (If the bottle is completely full, remove the stopper and discard about 20–30 ml of water; then replace the stopper and shake. This ensures thorough mixing.)

WHO 96603

**C.** With a sterile 10-ml pipette, inoculate 10 ml of the sample into each of five tubes containing 10 ml of presumptive broth (double strength). Add 50 ml of sample to a tube containing 50 ml of presumptive broth. It is advisable to shake the tubes gently to distribute the sample uniformly throughout the medium.

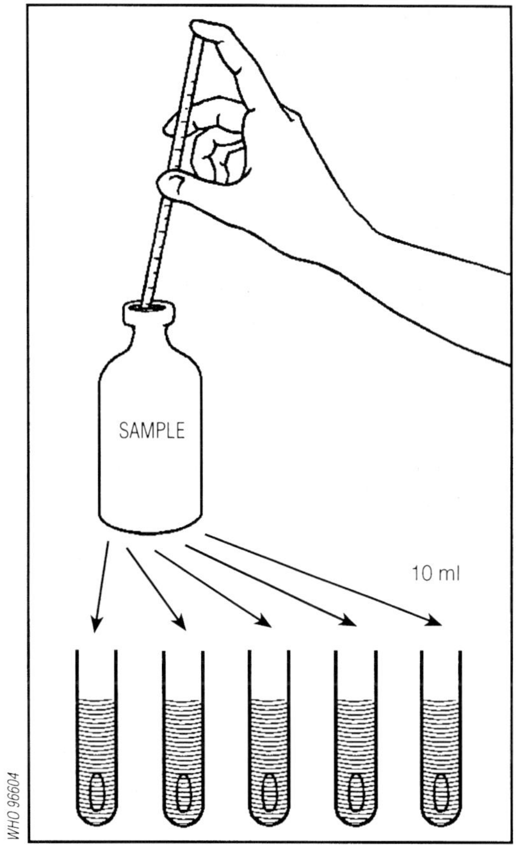


**D.** Incubate the tubes at 35 °C or 37 °C for 24 hours.

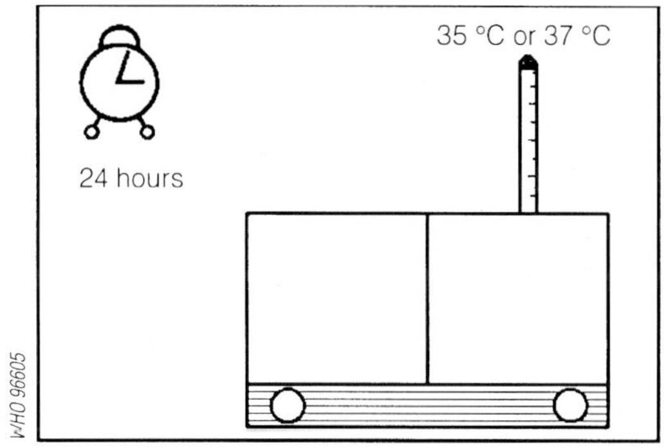

**E.** At the end of the 24-hour incubation period, examine each tube for the presence of gas. If present, gas can be seen in the Durham tube. If none is visible, gently shake the tube; if any effervescence (streams of tiny bubbles) is observed, the tube should be considered positive.

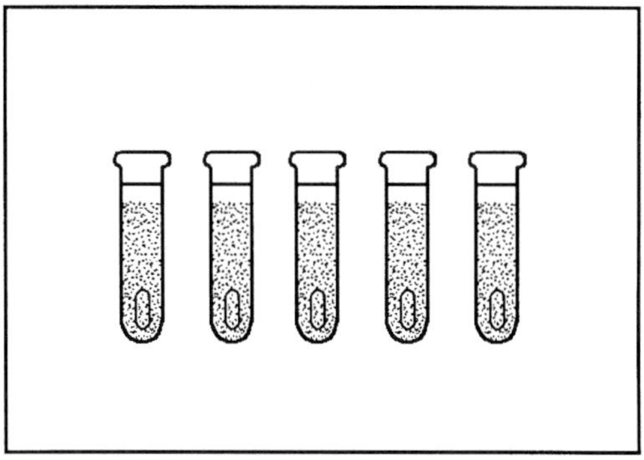

WHO 96606

**F.** Using a table like the one shown here, record the number of positive tubes after 24 hours.

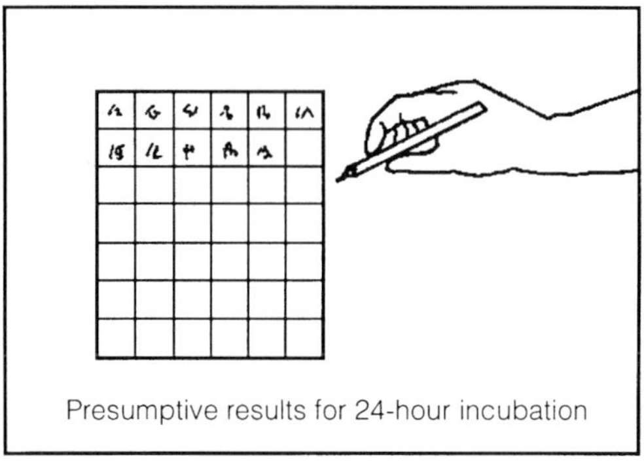


WHO 96607

**G.** Reincubate negative tubes for a further 24-hour period. At the end of this period, check the tubes again for gas production as in **E** above. Gas production at the end of either 24 or 48 hours' incubation is presumed to be due to the presence of coliforms in the sample.

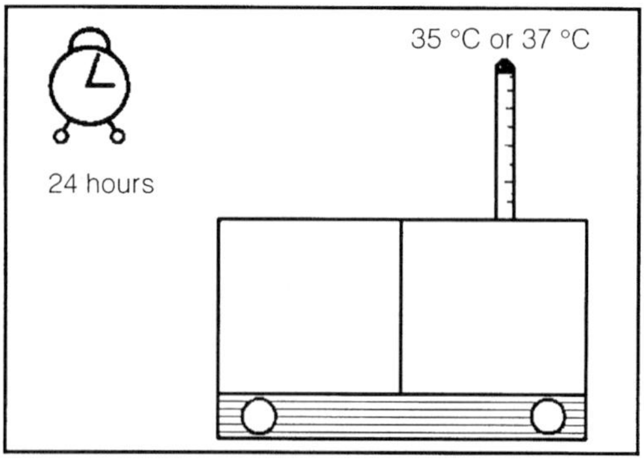


WHO 96608

**H.** Record the number of positive tubes after 48 hours.

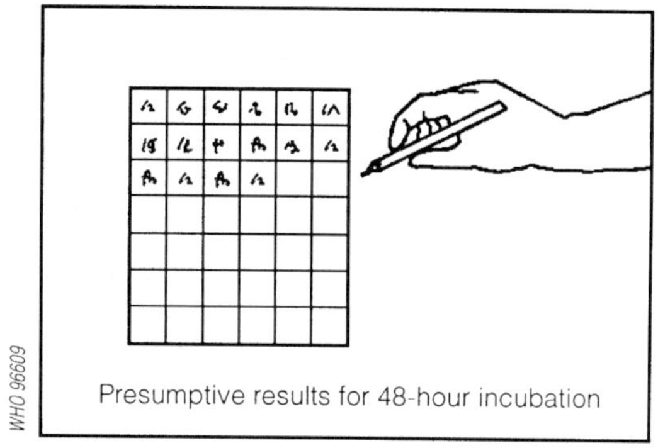


**I.** The confirmatory test should be carried out at the end of both the 24-hour and the 48-hour incubation. Using a sterile loop, transfer one or two drops from each presumptive positive tube into two tubes containing respectively confirmatory broth and tryptone water. (Sterilize the inoculation loop before each transfer by flaming and allow to cool.)

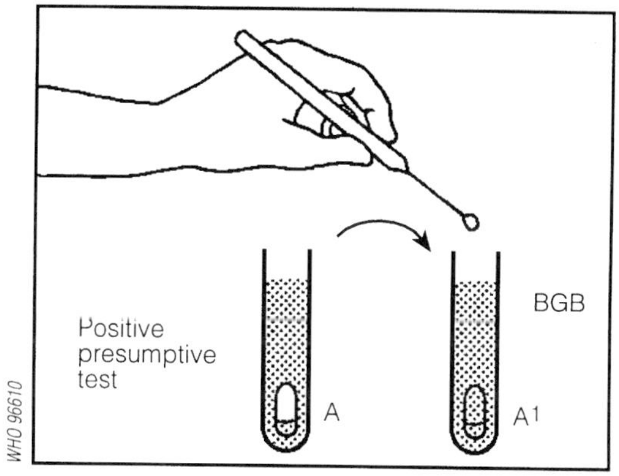


**J.** To confirm the presence of thermotolerant coliforms, incubate the subculture tubes from each presumptive positive tube for 24 hours at 44 ± 0.5 °C.

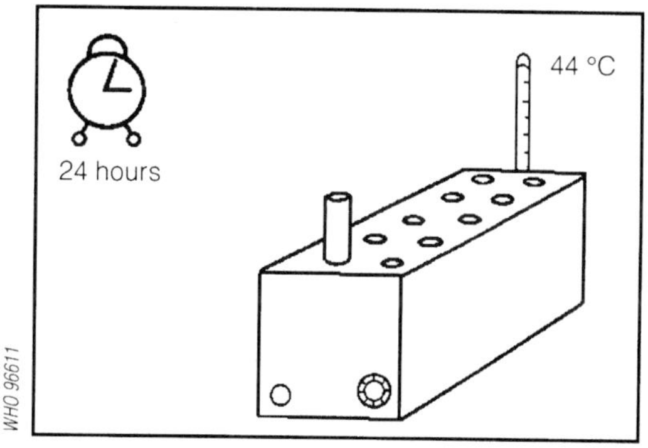

K. At the end of 24 hours' incubation, examine each broth tube for growth and the presence of gas in the Durham tube. Enter the results on the table as shown.

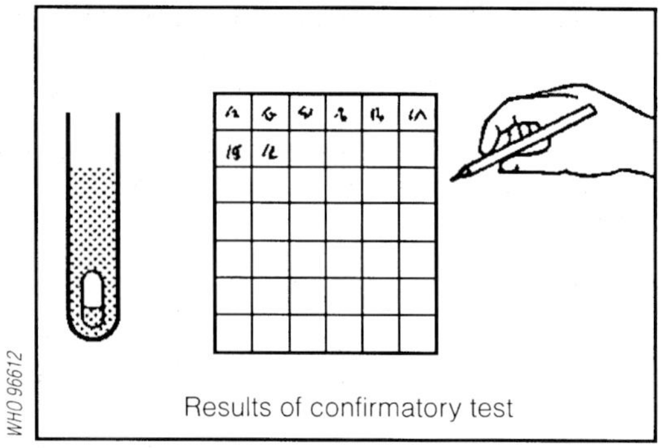

Results of confirmatory test

L. To each tube of tryptone water, add approximately 0.1 ml of Kovacs reagent (see Table A5.1, p. 192) and mix gently. The presence of indole is indicated by a red colour in the Kovacs reagent, forming a film over the aqueous phase of the medium.

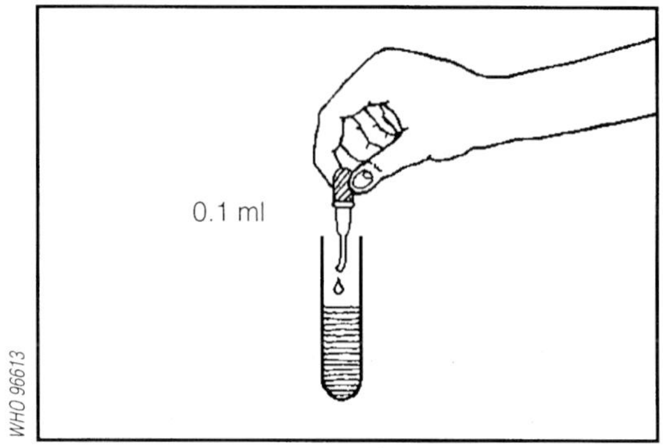


M. Confirmatory tests positive for indole, growth, and gas production show the presence of *E. coli.* Growth and gas production in the absence of indole confirms thermotolerant coliforms.

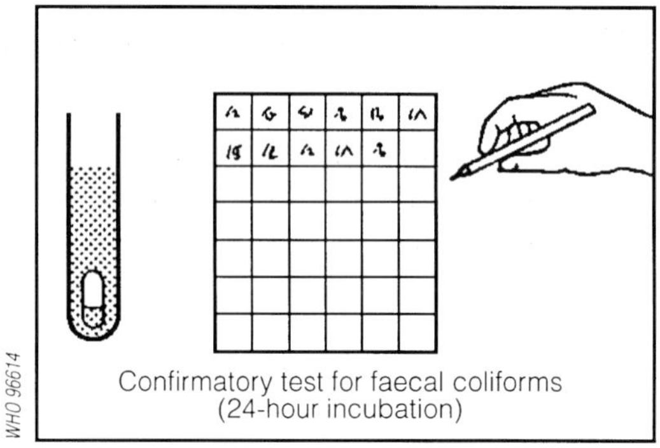

Confirmatory test for faecal coliforms (24-hour incubation)

***Table A5.2 MPN values per 100 ml of sample and 95% confidence limits for various combinations of positive and negative results (when one 50-ml and five 10-ml test portions are used)***

| No. of tubes giving a positive reaction | | MPN (per 100 ml) | 95% confidence limits | |
|---|---|---|---|---|
| 1 of 50 ml | 5 of 10 ml | | Lower | Upper |
| 0 | 0 | <1 | — | — |
| 0 | 1 | 1 | <1 | 4 |
| 0 | 2 | 2 | <1 | 6 |
| 0 | 3 | 4 | <1 | 11 |
| 0 | 4 | 5 | 1 | 13 |
| 0 | 5 | 7 | 2 | 17 |
| 1 | 0 | 2 | <1 | 6 |
| 1 | 1 | 3 | <1 | 9 |
| 1 | 2 | 6 | 1 | 15 |
| 1 | 3 | 9 | 2 | 21 |
| 1 | 4 | 16 | 4 | 40 |
| 1 | 5 | >18 | — | — |

### A5.6.2 Determination of MPN

For treated water, where one 50-ml and five 10-ml portions are inoculated, the MPN can be found from the test results by means of Table A5.2.

## A5.7 Application to polluted water (full method)

### A5.7.1 Procedure

The procedure to be used for the testing of water that is expected to be polluted, even though it may have been treated, is shown below and is essentially similar to that described in section A5.6, with the exception that several dilutions are used.

**A.** Arrange three rows of five tubes each in a test-tube rack. The tubes in the first row (F1) hold 10 ml of double-strength presumptive medium while the tubes in the second and third rows (F2, F3) contain 10 ml of single-strength presumptive medium.

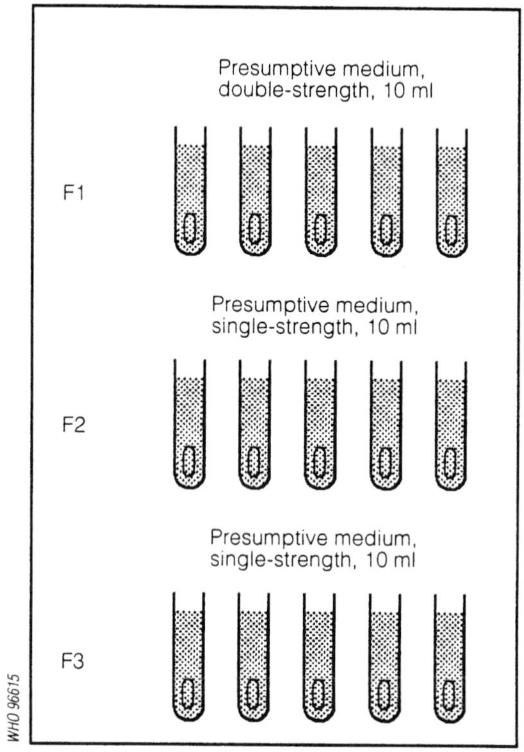


**B.** With a sterile pipette add 10 ml of sample to each of the five tubes in row F1.

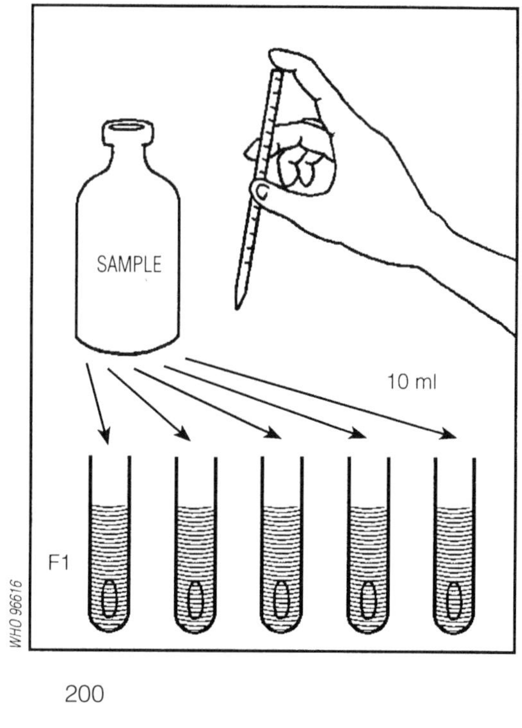

**C.** With a sterile pipette, add 1ml of sample to each of the five tubes in row F2.

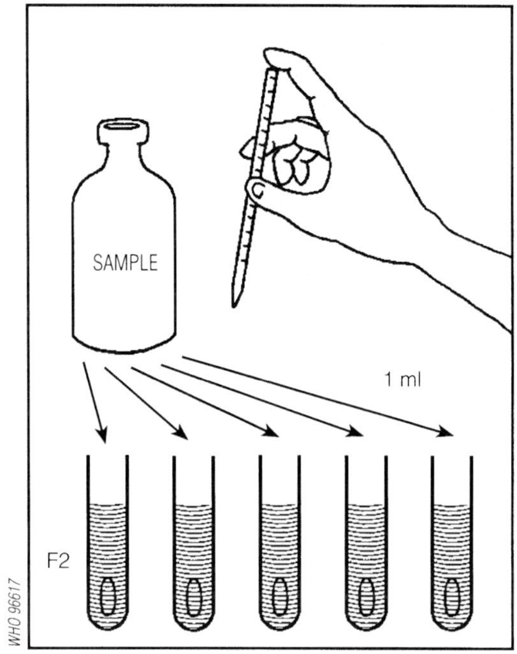

**D.** Prepare a 1 : 10 dilution of the sample by adding 1 ml of sample to 9 ml of dilution water (use a 1-ml sterile pipette). Recap the bottle containing the diluted sample and shake it vigorously.

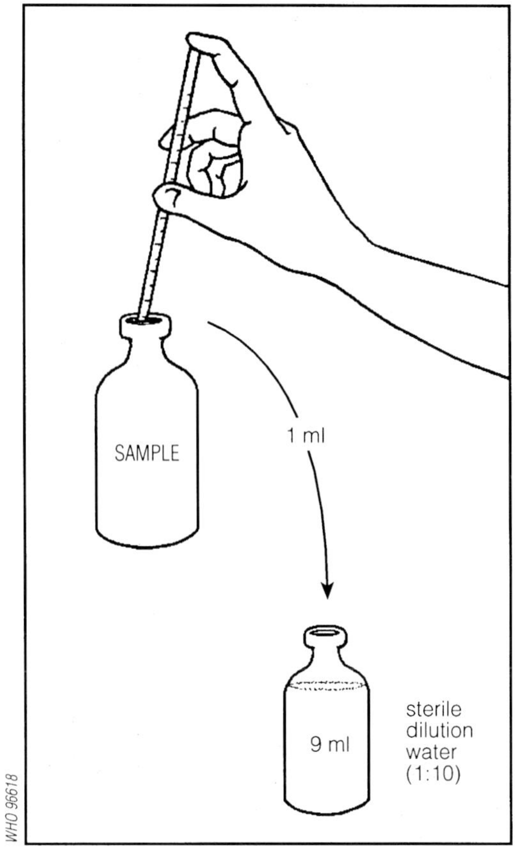

**E.** With another sterile pipette add 1 ml of the 1:10 dilution to each of the five tubes in row F3.

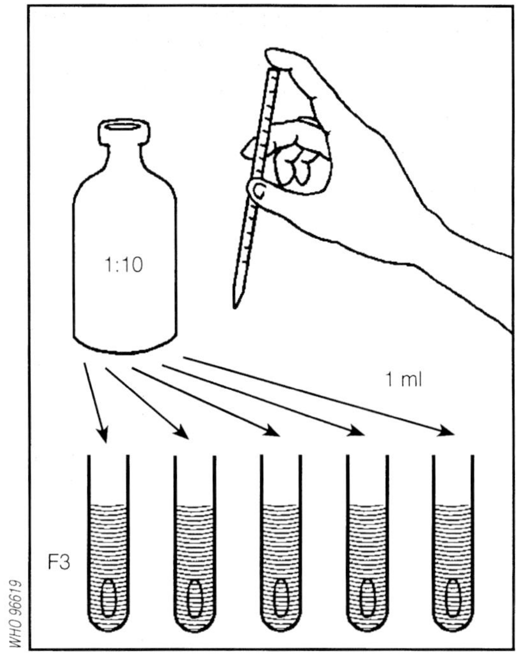


**F.** After gently shaking the tubes to mix the contents, incubate the rack with the 15 tubes at 35 °C or 37 °C for 24 hours. Then proceed in the same way as for *unpolluted* water.

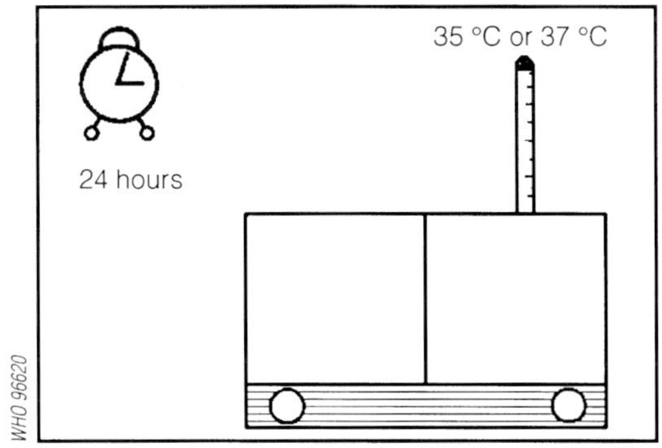


## A5.7.2 Determination of MPN

The MPN is found in a similar way to that described in section A5.6.2 but, because of the large number of tubes involved, a more complicated table—Table A5.3—must be used.

The following example shows how the results are obtained.

Suppose that, after confirmation of the presence of thermotolerant (faecal) coliforms, the following results are obtained:

— 5 positive tubes in row F1 (sample volume inoculated, 10 ml)
— 3 positive tubes in row F2 (sample volume inoculated, 1 ml)
— 1 positive tube in row F3 (sample volume inoculated, 0.1 ml).

***Table A5.3 MPN values per 100 ml of sample and 95% confidence limits for various combinations of positive and negative results (when five 10-ml, five 1-ml and five 0.1 ml test portions are used)***

| No. of tubes giving a positive reaction : | | | MPN (per 100 ml) | 95% confidence limits | |
|---|---|---|---|---|---|
| 5 of 10 ml | 5 of 1 ml | 5 of 0.1 ml | | Lower | Upper |
| 0 | 0 | 0 | <2 | <1 | 7 |
| 0 | 1 | 0 | 2 | <1 | 7 |
| 0 | 2 | 0 | 4 | <1 | 11 |
| 1 | 0 | 0 | 2 | <1 | 7 |
| 1 | 0 | 1 | 4 | <1 | 11 |
| 1 | 1 | 0 | 4 | <1 | 11 |
| 1 | 1 | 1 | 6 | <1 | 15 |
| 2 | 0 | 0 | 5 | <1 | 13 |
| 2 | 0 | 1 | 7 | 1 | 17 |
| 2 | 1 | 0 | 7 | 1 | 17 |
| 2 | 1 | 1 | 9 | 2 | 21 |
| 2 | 2 | 0 | 9 | 2 | 21 |
| 2 | 3 | 0 | 12 | 3 | 28 |
| 3 | 0 | 0 | 8 | 1 | 19 |
| 3 | 0 | 1 | 11 | 2 | 25 |
| 3 | 1 | 0 | 11 | 2 | 25 |
| 3 | 1 | 1 | 14 | 4 | 34 |
| 3 | 2 | 0 | 14 | 4 | 34 |
| 3 | 2 | 1 | 17 | 5 | 46 |
| 3 | 3 | 0 | 17 | 5 | 46 |
| 4 | 0 | 0 | 13 | 3 | 31 |
| 4 | 0 | 1 | 17 | 5 | 46 |
| 4 | 1 | 0 | 17 | 5 | 46 |
| 4 | 1 | 1 | 21 | 7 | 63 |
| 4 | 1 | 2 | 26 | 9 | 78 |
| 4 | 2 | 0 | 22 | 7 | 67 |
| 4 | 2 | 1 | 26 | 9 | 78 |
| 4 | 3 | 0 | 27 | 9 | 80 |
| 4 | 3 | 1 | 33 | 11 | 93 |
| 4 | 4 | 0 | 34 | 12 | 93 |
| 5 | 0 | 0 | 23 | 7 | 70 |
| 5 | 0 | 1 | 31 | 11 | 89 |
| 5 | 0 | 2 | 43 | 15 | 110 |
| 5 | 1 | 0 | 33 | 11 | 93 |
| 5 | 1 | 1 | 46 | 16 | 120 |
| 5 | 1 | 2 | 63 | 21 | 150 |
| 5 | 2 | 0 | 49 | 17 | 130 |
| 5 | 2 | 1 | 70 | 23 | 170 |
| 5 | 2 | 2 | 94 | 28 | 220 |
| 5 | 3 | 0 | 79 | 25 | 190 |
| 5 | 3 | 1 | 110 | 31 | 250 |
| 5 | 3 | 2 | 140 | 37 | 340 |
| 5 | 3 | 3 | 180 | 44 | 500 |

***Table A5.3 (continued)***

| No. of tubes giving a positive reaction : | | | MPN (per 100 ml) | 95% confidence limits | |
|---|---|---|---|---|---|
| 5 of 10 ml | 5 of 1 ml | 5 of 0.1 ml | | Lower | Upper |
| 5 | 4 | 0 | 130 | 35 | 300 |
| 5 | 4 | 1 | 170 | 43 | 490 |
| 5 | 4 | 2 | 220 | 57 | 700 |
| 5 | 4 | 3 | 280 | 90 | 850 |
| 5 | 4 | 4 | 350 | 120 | 1000 |
| 5 | 5 | 0 | 240 | 68 | 750 |
| 5 | 5 | 1 | 350 | 120 | 1000 |
| 5 | 5 | 2 | 540 | 180 | 1400 |
| 5 | 5 | 3 | 920 | 300 | 3200 |
| 5 | 5 | 4 | 1600 | 640 | 5800 |
| 5 | 5 | 5 | >1800 | — | — |

The results can thus be coded as 5–3–1; they represent the confirmatory test for thermotolerant coliforms. Table A5.3 indicates that a coded result of 5–3–1 (5 × 10 ml positive, 3 × 1 ml positive, 1 × 0.1 ml positive) gives an MPN value of 110, i.e. the water sample contains an estimated 110 coliforms per 100 ml.

Next, consider an example of heavily polluted water. The procedure outlined above may give a coded result of 5–5–5. Such a result does not give a definite MPN value. When such heavy contamination is suspected it is usual to inoculate more than three dilutions in a series of factors of 10. This series of 10-fold dilutions should be made in such a way that a negative result is likely for at least the highest dilution incubated. If 5 × 1.0 ml, 5 × 0.1 ml, 5 × 0.01 ml, and 5 × 0.001 ml are initially inoculated and a confirmed coded result of 5–5–4–1 is obtained, only three of these results should then be used to obtain the MPN value from Table A5.3. These should be selected by choosing the smallest sample volume (in this case, 0.1 ml) for which all the tubes give a positive result, and the two next succeeding higher dilutions. The coded result of these three volumes is then used to obtain the MPN value from Table A5.3. In the above example, the result 5–4–1 would be chosen, representing volumes of 0.1, 0.01, and 0.001 ml of the sample. The MPN value obtained from Table A5.3 should be multiplied by 100 to obtain the MPN for this particular sample (see below); in this case, the result is 17 000 per 100 ml.

Sometimes the laboratory worker may find it difficult to determine the multiplying factor to be used to obtain the appropriate MPN for the sample tested. A simple way to determine the MPN is to divide the MPN value obtained from Table A5.3 by the sample volume represented by the middle number in the chosen code. For example, consider a chosen code of 5–2–0, in which the 2 represents a sample volume of 0.01 ml (see Table A5.4). From Table A5.3, MPN for a code of 5–2–0 is 49. The MPN value for the sample tested will therefore be:

$$(49/0.01) = 49 \times 100 = 4900.$$

**Table A5.4 Example of multiplying factors for determination of the MPN for different dilutions of sample**

| Example | No. of tubes giving a positive reaction: 5 of 1 ml | 5 of 0.1 ml | 5 of 0.01 ml | 5 of 0.001 ml | 5 of 0.0001 ml | Coded result chosen | Multiplying factor for MPN |
|---|---|---|---|---|---|---|---|
| 1 | 5 | 5 | 2 | 0 | 0 | 5–2–0 | 100 |
| 2 | 5 | 5 | 4 | 1 | 0 | 5–4–1 | 100 |
| 3 | 5 | 3 | 0 | 0 | 0 | 5–3–0 | 10 |
| 4 | 5 | 5 | 5 | 3 | 1 | 5–3–1 | 1000 |
| 5 | 0 | 1 | 0 | 0 | 0 | 0–1–0 | 10 |

**Table A5.5 MPN values per 100 ml of sample and 95% confidence limits for various combinations of positive and negative results (when three 10-ml, three 1-ml, and three 0.1-ml test portions are used)**

| No. of tubes giving a positive reaction: 3 of 10 ml | 3 of 1 ml | 3 of 0.1 ml | MPN (per 100 ml) | 95% confidence limits: Lower | Upper |
|---|---|---|---|---|---|
| 0 | 0 | 1 | 3 | <1 | 9 |
| 0 | 1 | 0 | 3 | <1 | 13 |
| 0 | 0 | 0 | 4 | <1 | 20 |
| 1 | 0 | 1 | 7 | 1 | 21 |
| 1 | 1 | 0 | 7 | 1 | 23 |
| 1 | 1 | 1 | 11 | 3 | 36 |
| 1 | 2 | 0 | 11 | 3 | 36 |
| 2 | 0 | 0 | 9 | 1 | 36 |
| 2 | 0 | 1 | 14 | 3 | 37 |
| 2 | 1 | 0 | 15 | 3 | 44 |
| 2 | 1 | 1 | 20 | 7 | 49 |
| 2 | 2 | 0 | 21 | 4 | 47 |
| 2 | 2 | 1 | 28 | 10 | 149 |
| 3 | 0 | 0 | 23 | 4 | 120 |
| 3 | 0 | 1 | 39 | 7 | 130 |
| 3 | 0 | 2 | 64 | 15 | 379 |
| 3 | 1 | 0 | 48 | 7 | 210 |
| 3 | 1 | 1 | 75 | 14 | 230 |
| 3 | 1 | 2 | 120 | 30 | 380 |
| 3 | 2 | 0 | 93 | 15 | 380 |
| 3 | 2 | 1 | 150 | 30 | 440 |
| 3 | 2 | 2 | 210 | 35 | 470 |
| 3 | 3 | 0 | 240 | 36 | 1300 |
| 3 | 3 | 1 | 460 | 71 | 2400 |
| 3 | 3 | 2 | 1100 | 150 | 4800 |

Examples are given in Table A5.4 of the factors to be used to multiply the MPN value found in order to obtain the appropriate MPN for different dilutions.

## A5.8 Application to polluted water: "shorter method"

The procedure for the shorter method is almost identical to that described in section A5.7, with the single difference that only three tubes of each sample volume are inoculated, instead of five. This requires the use of a different table—Table A5.5—for determining the MPN.

## A5.9 Direct thermotolerant coliform method

If *unchlorinated* water from small-community water supplies is tested and only the number of thermotolerant coliforms is of interest, a direct multiple-tube method can be used. This is recommended for use where the total coliform result is not of great significance, e.g. in small-community supplies in developing countries or where space, time, or facilities are limited. The method is based on the normal MPN procedure, but the tubes are incubated directly in a water-bath at 44.5 ± 0.2 °C, without previously incubating at 35 or 37 °C for 24 hours and testing for total coliforms.

The procedure is similar to that described for the examination of polluted water, except that MacConkey broth is used as the presumptive medium. Prepare 15 tubes of sample and medium, as described on pp. 199–203, and then proceed as shown below.

**A.** After gently shaking the tubes to mix the contents, incubate the 15 tubes at 44 °C for 24 hours.

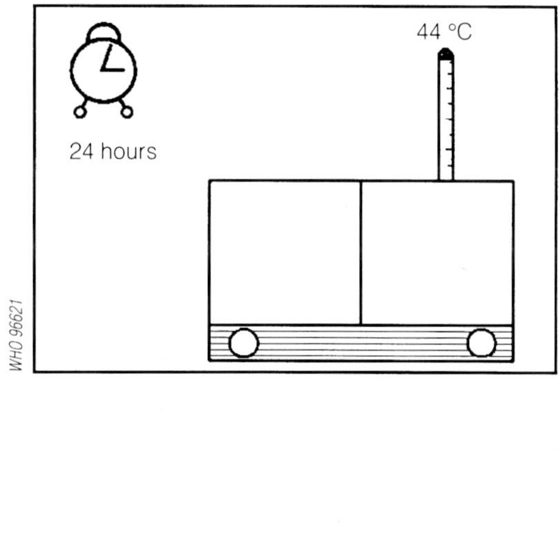

**B.** Observe each tube for the presence of gas and enter the number of positive tubes after 24 hours in the appropriate table.

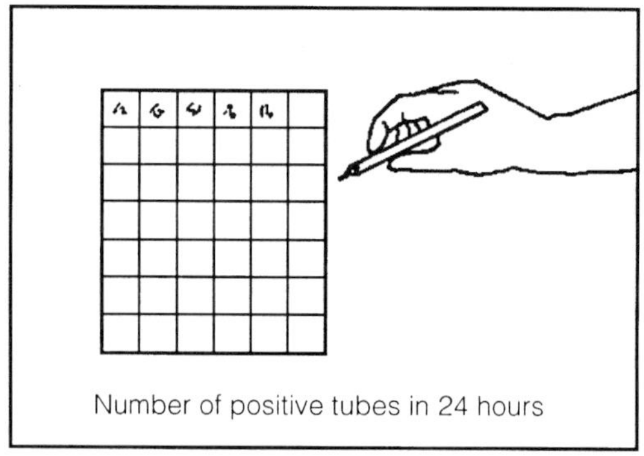


WHO 96622

**C.** Negative tubes should be reincubated for a further 24-hour period, after which they should be observed for the presence of gas.

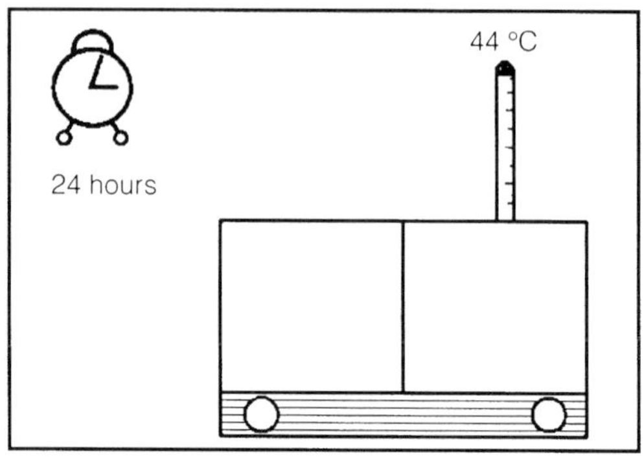


WHO 96621

**D.** Confirm the presumptive results after 24 and 48 hours by transferring a loopful of broth to a confirmatory broth and incubating at 44 °C for 24 hours.

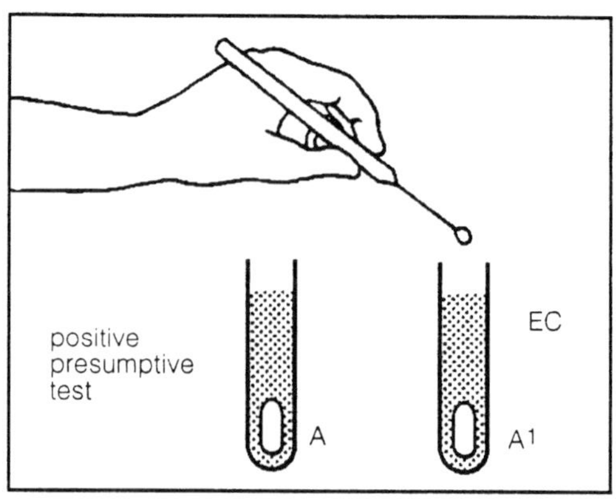


WHO 96623

E. The presence of thermotolerant coliforms is confirmed if gas is present in the confirmatory broth after 24 hours at 44 °C. Determine the MPN from Table A5.3 as before.

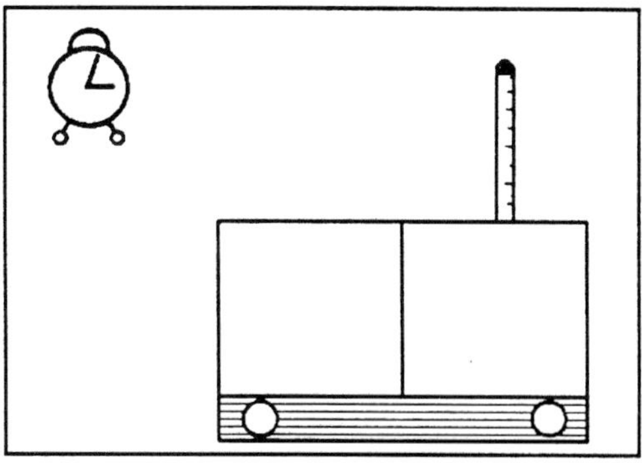


## A5.10 Selection of tubes for confirmatory test

Any bacteriological analysis should always include the confirmatory test. If only five 10-ml portions are tested, the confirmatory test for coliforms and thermotolerant coliforms must be carried out on all tubes showing gas production. However, if the inoculation involved five (or three) tubes for each of, or more than, three sample volumes (e.g. 10, 1.0, 0.1, 0.01, and 0.001 ml), it is not necessary to carry out confirmatory tests on all the positive tubes.

If all five (or three) tubes of two or more consecutive dilutions are positive, the set of tubes should be selected that presents the smallest sample volume for which all the tubes are positive. The confirmatory test should be carried out on all these tubes and on all the positive tubes corresponding to subsequent and lower volumes. The following example should help to illustrate this procedure. After 24 hours' incubation, five tubes with 10 ml, five with 1.0 ml, five with 0.1 ml, four with 0.01 ml, and one with 0.001 ml gave positive results. Thus the confirmatory test should be carried out on the positive tubes initially inoculated with 0.1, 0.01, and 0.001 ml of sample.

## A5.11 Record forms

The analysis of a given sample will provide several results. The form drawn up for recording these results, although it should not be complicated, must be completed. The completed form should contain the data on the sampling, which will also serve to identify the samples, those entered on the sample dispatch form, and data on the bacteriological analysis itself. A suggested comprehensive form is shown in Fig. A5.1. Once the analysis is completed, the laboratory carrying out the work should record the results obtained in a standardized form (protocol); this should follow the recommendations given in Chapter 3. The protocol can be a very simple report, which records the sample identification information together with the result of the analysis and the appropriate classification of the water. An example of such a protocol is shown in Fig. A5.2.

## *Fig. A5.1 Suggested comprehensive form for recording results of analyses by the multiple-tube method*

| Sample identification | | | | | Sender | Data and time of collection | Date and time of first analysis | Residual free chlorine (mg/litre) | Sample volume inoculated (ml) | No. of tubes inoculated | No. of positive tubes in presumptive test 35 or 37 °C | | | No. of positive tubes in confirmatory test | | MPN | |
|---|---|---|---|---|---|---|---|---|---|---|---|---|---|---|---|---|---|
| Sample no. | Com-munity | Sample type | Place | Source | | | | | | | 24 hours | 48 hours | Total | 35 or 37 °C, 48 h | 44 °C, 24 h | Total coli-forms | Faecal coli-forms |
| | | | | | | | | | | | 10 | | | | | | |
| | | | | | | | | | | | 1.0 | | | | | | |
| | | | | | | | | | | | 0.1 | | | | | | |
| | | | | | | | | | | | | | | | | | |
| | | | | | | | | | | | | | | | | | |
| | | | | | | | | | | | 10 | | | | | | |
| | | | | | | | | | | | 1.0 | | | | | | |
| | | | | | | | | | | | 0.1 | | | | | | |
| | | | | | | | | | | | | | | | | | |
| | | | | | | | | | | | | | | | | | |
| | | | | | | | | | | | 10 | | | | | | |
| | | | | | | | | | | | 1.0 | | | | | | |
| | | | | | | | | | | | 0.1 | | | | | | |
| | | | | | | | | | | | | | | | | | |
| | | | | | | | | | | | | | | | | | |
| | | | | | | | | | | | 10 | | | | | | |
| | | | | | | | | | | | 1.0 | | | | | | |
| | | | | | | | | | | | 0.1 | | | | | | |
| | | | | | | | | | | | | | | | | | |
| | | | | | | | | | | | | | | | | | |
| | | | | | | | | | | | 10 | | | | | | |
| | | | | | | | | | | | 1.0 | | | | | | |
| | | | | | | | | | | | 0.1 | | | | | | |
| | | | | | | | | | | | | | | | | | |
| | | | | | | | | | | | | | | | | | |

***Fig. A5.2 Suggested protocol for results of bacteriological analysis***

WATER-QUALITY
CONTROL
PROGRAMME

BACTERIOLOGICAL
WATER
ANALYSIS

[. . . . . . . . . . . . . . . . . . . . . . . . . . . . . . Authority]

COMMUNITY:

SAMPLE NO. . . . . . . . .

SAMPLE SITE:

PLACE:

SOURCE:

SENDER:

DATE OF SAMPLING ______/____/______ TIME:

DATE OF ANALYSIS ______/____/______ TIME:

RESIDUAL FREE CHLORINE [     ] mg/litre

RESULTS

TOTAL COLIFORMS . . . . . . . . . . . . /100 ml

FAECAL COLIFORMS . . . . . . . . . . . /100 ml

WATER BACTERIOLOGICALLY

GOOD – BAD

..................................................
Laboratory Technician

..................................................
Chief (Signed)

WHO 96624

Annex 6

# Membrane filtration method for thermotolerant (faecal) coliforms

## A6.1 Principle

In contrast to the multiple-tube method, the membrane-filtration method gives a direct count of total coliforms and thermotolerant coliforms present in a given sample of water. The method is based on the filtration of a known volume of water through a membrane filter consisting of a cellulose compound with a uniform pore diameter of 0.45 or 0.2 μm; the bacteria are retained on the surface of the membrane filter. When the membrane containing the bacteria is incubated in a sterile container at an appropriate temperature with a selective differential culture medium, characteristic colonies of thermotolerant coliforms develop, which can be counted directly.

## A6.2 Volume of water sample for filtration

Since the filtration area is relatively small, it can support the growth of only a limited number of colonies: the optimum number is between 20 and 80, with a maximum of 200. If this figure is exceeded, very small atypical colonies or superimposed colonies may develop, or there may be growth inhibition due to overpopulation. The choice of the volume of sample to be filtered will depend on the type of water. Examples of typical volumes are provided in Table 4.3 (p. 61).

## A6.3 Equipment and glassware

In addition to the basic equipment and glassware used in the multiple-tube method (see Annex 5), the following items are needed for the membrane-filtration technique:

- *Membrane-filtration apparatus*: including an electric or hand-powered vacuum pump, a vacuum flask (e.g. an Erlenmeyer side-arm flask), and a filter support.
- *Reusable Petri dishes*: made from glass or metal (disposable plastic Petri dishes may also be used).
- *Blunt-ended forceps*: for picking up membrane filters.
- *Reusable (autoclavable) bottles*: for culture media (e.g. 25-ml polypropylene bottles).

- *A magnifying lens*: with ×4 or ×5 magnification for examining and counting the colonies on the membrane filters.
- *A boiling bath/pan*: if filtration apparatus is to be disinfected in boiling water between analyses.
- *Sterile pipettes*: 1 ml and 10 ml.
- *A graduated cylinder*: 100 ml.

In addition to the consumables needed for the MPN, the following are required:

- *Membrane filters*: 47–50 mm in diameter, with a pore diameter of 0.45 μm. Singly packed, presterilized membrane filters are very convenient. Unsterilized membrane filters can also be used, however, and should be wrapped in paper packets in convenient numbers (depending on the number of water samples to be tested). These can then be sterilized in the autoclave and dried by rapid exhaustion of the steam.
- *Nutrient absorbent pads*: filter-paper discs about 1 mm thick, with the same diameter as the membrane filters.
- *Culture media*: different types are available (see section A6.4).
- *Wax pencils*: for labelling Petri dishes.
- *Polythene bags*: for wrapping Petri dishes if a dry incubator is used, to prevent drying of the sample and media.

## A6.4 Culture media and dilution water

Various media can be used for the examination of coliform organisms by the membrane-filtration method. Of these, lactose Tergitol[1] agar, lactose TTC Tergitol[1] agar, and membrane lauryl sulfate lactose broth may be used for coliform organisms at 35 or 37 °C and for thermotolerant coliform organisms at 44 °C or 44.5 °C. Membrane faecal coliform (MFC) broth should be used only at 44 or 44.5 °C for thermotolerant coliform counts. Although the use of all these media for the detection of presumptive coliform organisms is based on the fermentation of lactose, the characteristic reaction varies with each medium, as shown in Table A6.1.

Although it is possible to prepare the media from the basic ingredients, this may be impractical in a small laboratory. The use of dehydrated media is therefore recommended. The media can be prepared as a broth and used together with nutrient absorption pads, or as solid agar plates. The broths may be solidified by the addition of 1.2–1.5% agar before boiling.

## A6.5 Procedure

The procedure generally used is described here, but different types of filtration units and equipment exist.

[1] Tergitol 7 is an example of a suitable product available commercially. This information is given for the convenience of the user and does not constitute an endorsement of this product by WHO.

**Table A6.1 Colony characteristics following analysis by the membrane-filtration method[a]**

| Medium[b] | Colony characteristics | |
|---|---|---|
| | **Total coliforms at 35/37 °C** | **Thermotolerant coliforms at 44/45.5 °C** |
| Lactose TTC[c] agar with Tergitol 7 | Yellow, orange or brick-red coloration with yellow central halo in the medium under the membrane | As for total coliforms at 35/37 °C |
| Lactose agar with Tergitol 7 | Yellow central halo in the medium under the membrane | As for total coliforms at 35/37 °C |
| Membrane-enriched Teepol broth | Yellow colour extending on to the membrane | As for total coliforms at 35/37 °C |
| Membrane lauryl sulfate broth | Yellow colour extending on to the membrane | As for total coliforms at 35/37 °C |
| Endo agar or broth | Dark red colour with golden-green metallic sheen | — |
| LES–Endo agar | Dark red colour with golden-green metallic sheen | — |
| Membrane faecal coliform (MFC) broth | — | Blue colonies |

[a] Adapted from ISO 9308-1: 1990, Detection and enumeration of coliform organisms, thermotolerant coliform organisms and presumptive *Escherichia coli*—Part 1: Membrane filtration method.
[b] Tergitol 7 and Teepol are examples of suitable products available commercially. This information is given for the convenience of the user and does not constitute an endorsement of these products by ISO or WHO.
[c] 2,3,5-Triphenyltetrazolium chloride.

**A.** Connect the Erlenmeyer (side-arm) flask to the vacuum source (turned off) and place the porous support in position. If an electric pump is used, it is advisable to put a second flask between the Erlenmeyer flask and the vacuum source; this second flask acts as a water trap, and thus protects the electric pump.

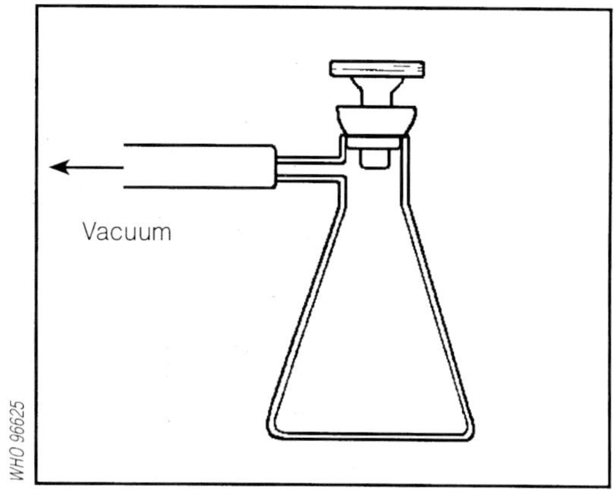


**B.** Open a sterile Petri dish and place a sterile absorbent pad in it.

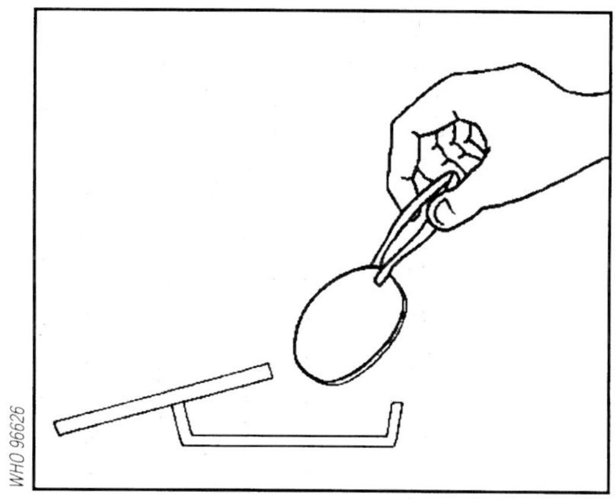


**C.** Add broth medium to saturate the pad; remove excess broth.

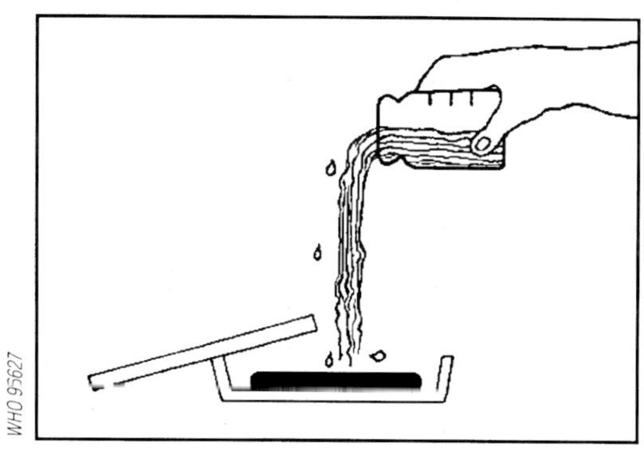

**D.** Assemble the filtration unit by placing a sterile membrane filter on the porous support, using forceps sterilized by flaming.

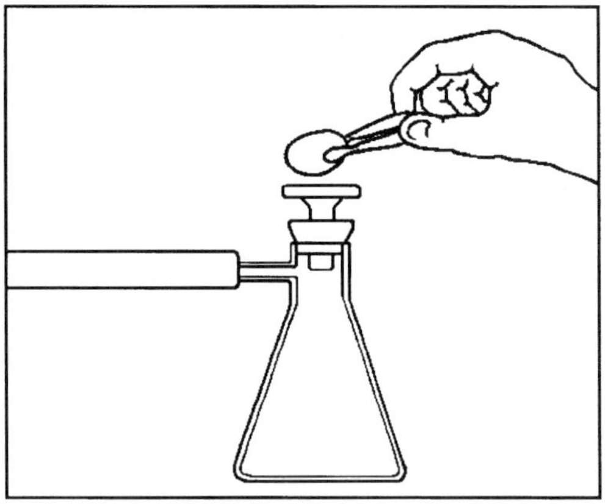

WHO 96628

E. Place the upper container in position and secure it. (The type of clamp used will depend on the type of equipment.)

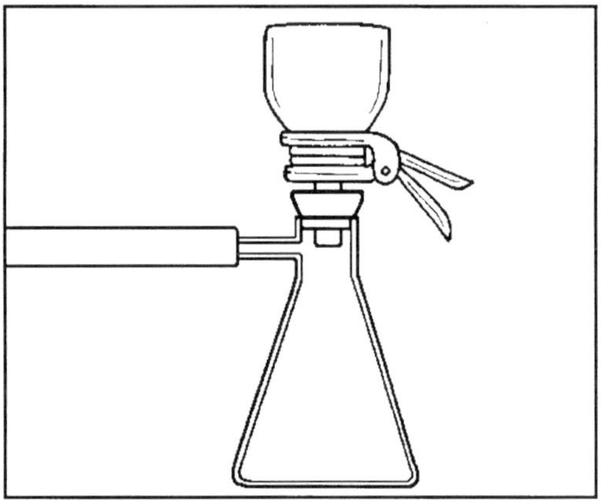

WHO 96629

F. Pour the volume of sample chosen as optimal for the type of water (see Table 4.3, p. 61), into the upper container. If the test sample is less than 10 ml, at least 20 ml of sterile dilution water should be added to the top container before filtration. Apply the vacuum.

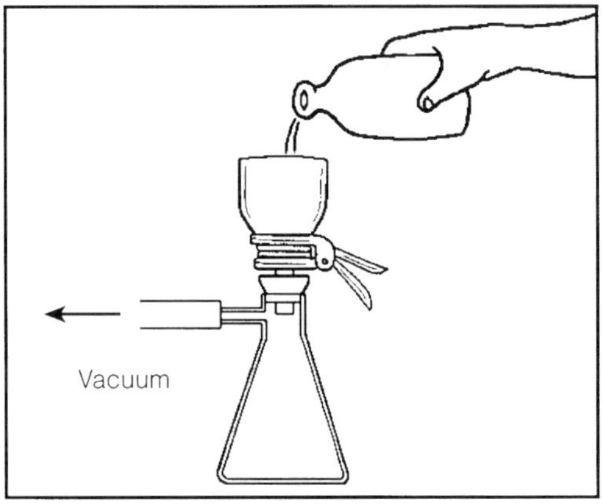


WHO 96630

**G.** Take the filtration unit apart and, using the sterile forceps, place the membrane filter in the Petri dish on the pad with the grid side up. Make sure that no air bubbles are trapped between the pad and the filter.

WHO 96631

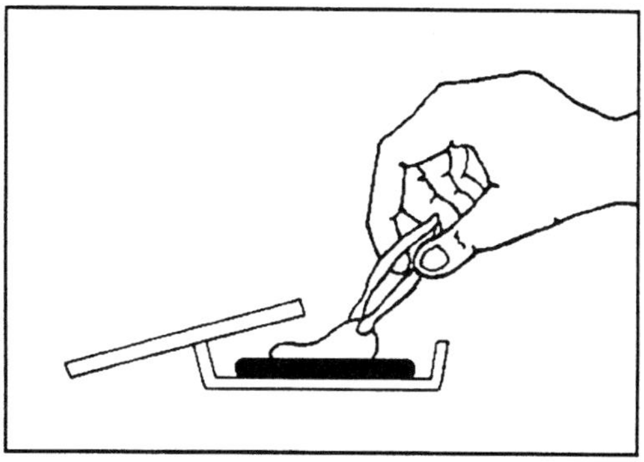

**H.** Leave the Petri dish at room temperature or at 35 or 37 °C for 2–4 hours, for resuscitation of stressed microbes.

WHO 96632

**I.** Place the dishes in an incubator at 44 ± 0.5 °C for 18–24 hours with 100% humidity. Alternatively, tight-fitting or sealed Petri dishes may be placed in waterproof plastic bags for incubation.

WHO 96633

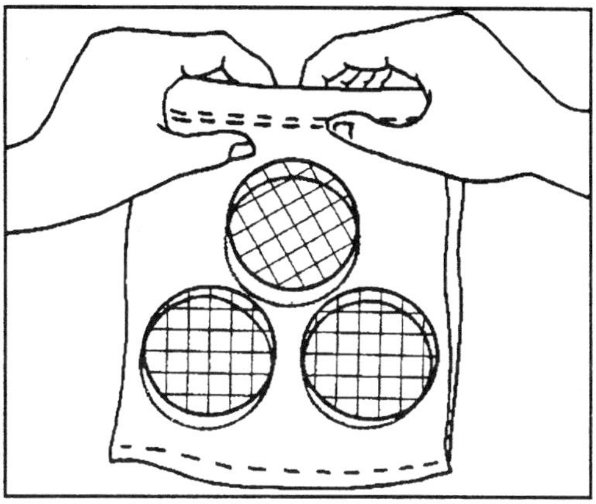

**J.** Submerge the bags in a water-bath maintained at 44 ± 0.5 °C for 18–24 hours. The plastic bags must be below the surface of the water throughout the incubation period. They can be held down by means of a suitable weight, e.g. a metal rack.

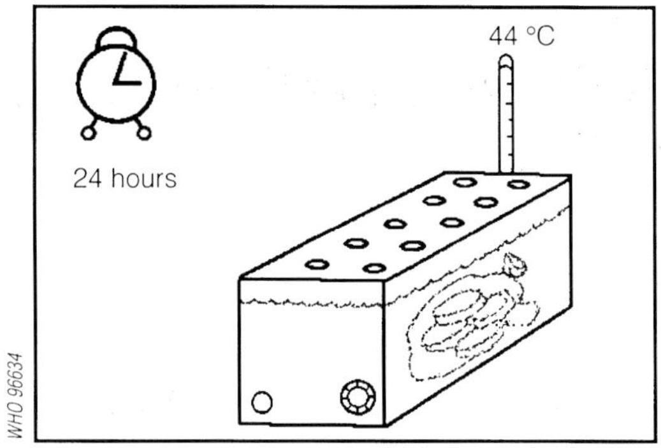


The colonies of thermotolerant coliform bacteria should be identified from their characteristics on the medium used. The number of thermotolerant coliforms per 100 ml is then given by:

Thermotolerant coliforms per 100 ml

$$= \frac{\text{no. of thermotolerant coliform colonies counted}}{\text{no. of ml of sample filtered}} \times 100$$

Annex 7

# Field test method for thermotolerant coliforms

The field test method for thermotolerant coliforms involves the following:

**A.** Flame-sterilize the tips of blunt-ended forceps and allow to cool between successive manipulations of the filters.

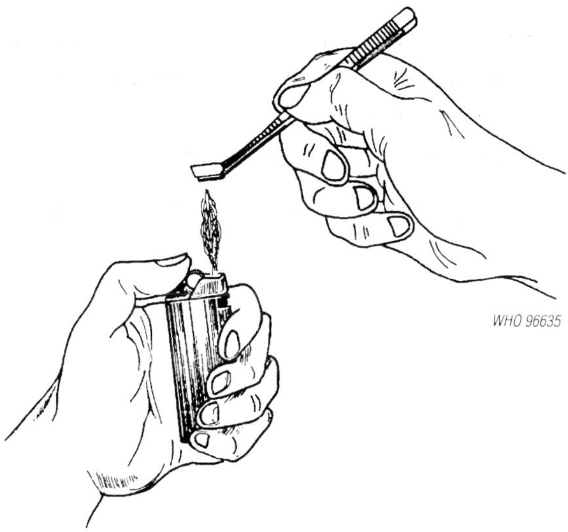


**B.** Place a sterile absorbent pad in a sterile Petri dish.

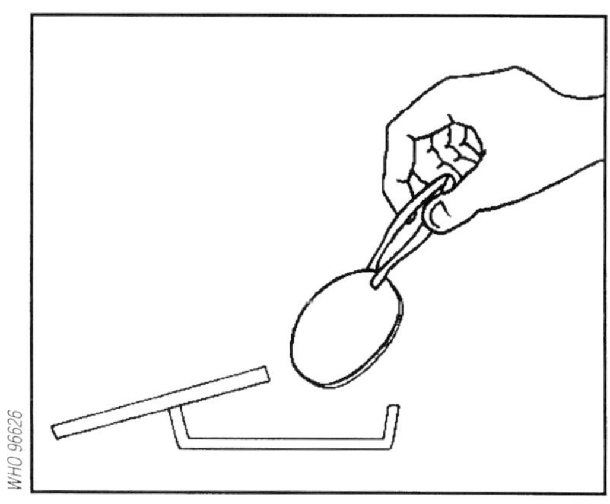

C. Add broth medium to saturate the pad and remove the excess broth.

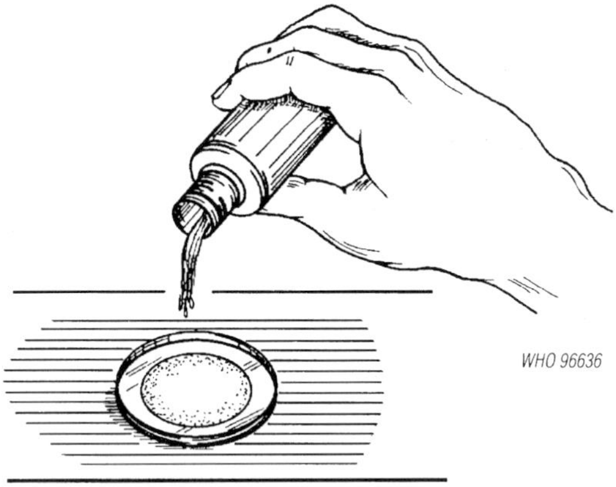

WHO 96636

D. Sterilize the filter apparatus and assemble by placing a sterile filter membrane on the membrane support.

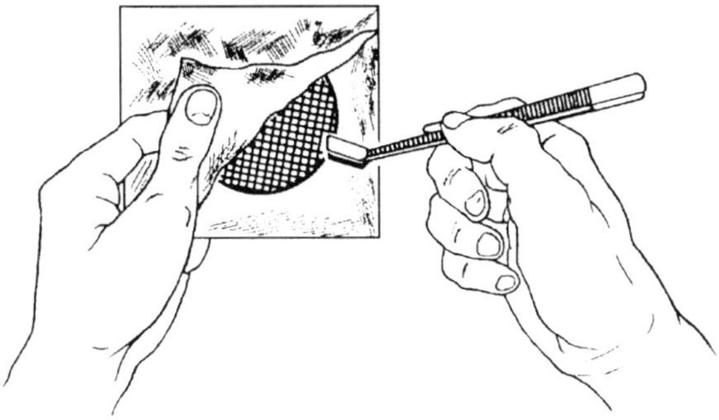

WHO 96637

E. Mix the sample thoroughly by inverting the sample bottle several times, and put the volume to be tested into the previously sterilized filtration apparatus. The appropriate volume of sample should be selected in accordance with the type of water being tested (see Table 4.3, p. 61).

WHO 96638

F. Apply a vacuum to the filter apparatus to draw the sample through the filter membrane. Disconnect the vacuum and dismantle the apparatus.

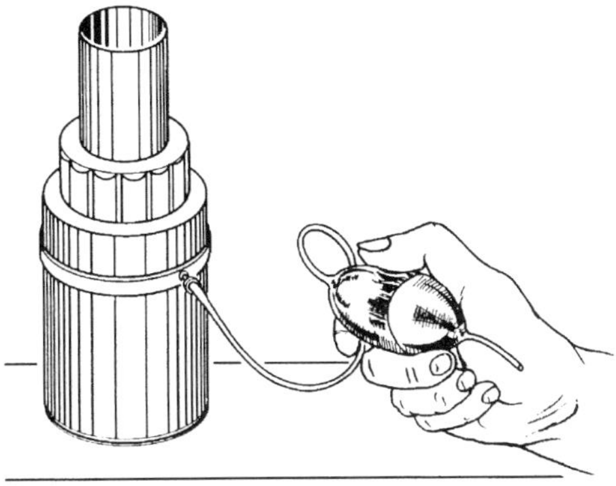

WHO 96639

**G.** Using sterile forceps, remove the membrane filter from the filter apparatus and transfer it to the nutrient pad in the Petri dish. Lower the membrane, grid side uppermost, carefully onto the nutrient pad, making sure that no air bubbles are trapped between the pad and the filter.

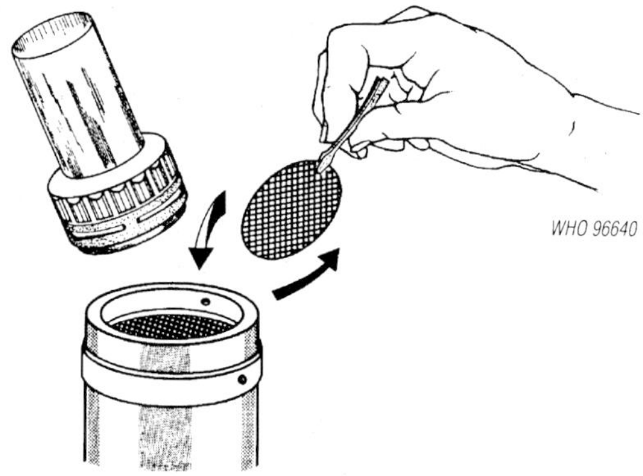

WHO 96640

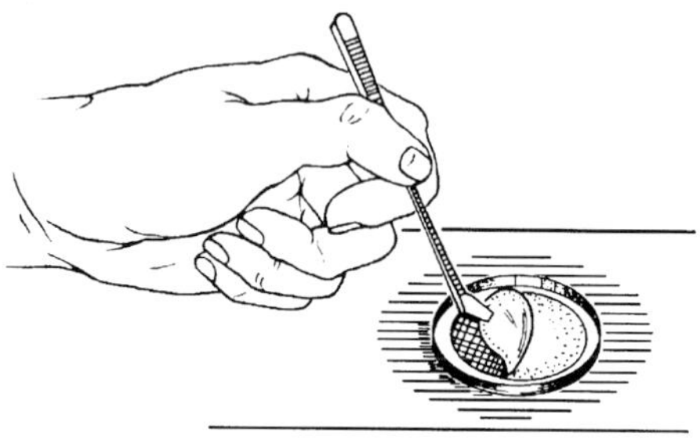

WHO 96641

**H.** Replace the lid on the Petri dish and label with the sample identification code using a wax pencil or waterproof pen.

WHO 96642

I. Incubate the Petri dish at ambient temperature for 2–4 hours to allow stressed bacteria to resuscitate.

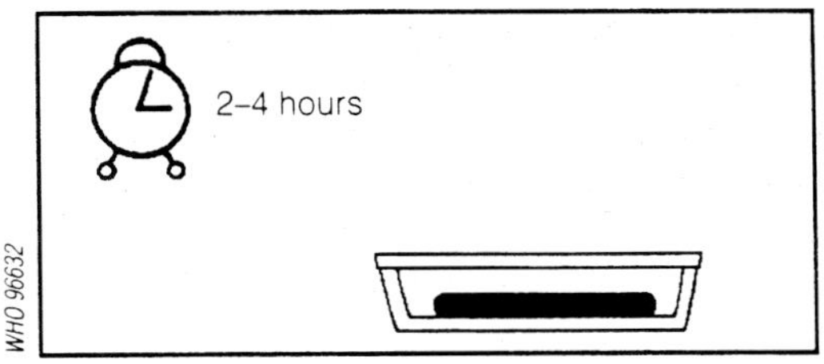


J. Incubate the Petri dish at the selected temperature for 18–24 hours.

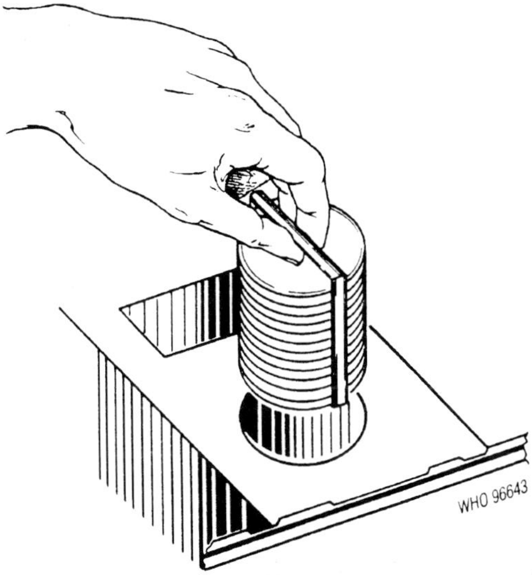


K. Following incubation, count all colonies with a morphology typical of the bacterium and the medium used. Calculate and express the result in colony-forming units (CFU) per 100 ml of sample.

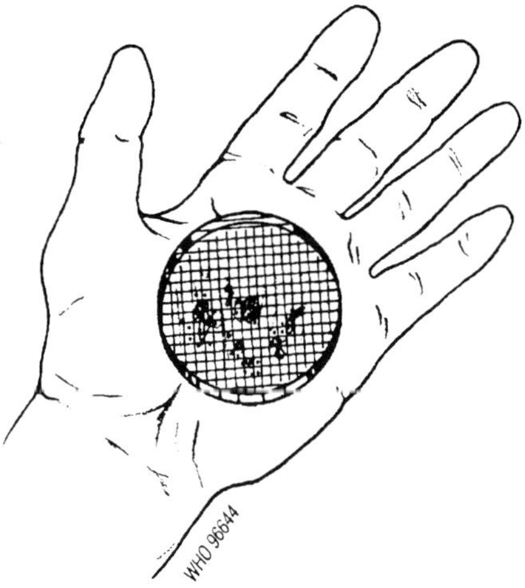

Annex 8

# Presence–absence test for total coliform bacteria

Presence–absence tests may sometimes be appropriate where positive results are known to be rare. They are not quantitative and, as their name suggests, they indicate only the presence or absence of the indicator sought. Such results are of little use in countries or situations where contamination is common and the purpose of the analysis is then to determine the degree of contamination rather than simply to indicate its presence. Thus presence–absence tests are *not recommended* for use in the analysis of surface waters, untreated small-community supplies, or larger water supplies in countries where operation and maintenance difficulties may occasionally occur.

Before a decision is taken to use the presence–absence test for the analysis of a water source, the results obtained by the test should be compared with those obtained with a recognized, quantitative method of analysis. Approximately 100 samples should be examined by both methods.

## A8.1 Preparation of medium

The constituents of the medium used for the presence–absence test for coliform bacteria are as follows:

| | |
|---|---|
| lactose broth (dehydrated) | 13.0 g |
| lauryl tryptose broth (dehydrated) | 17.5 g |
| bromocresol purple | 0.0085 g |
| distilled water | 1 litre |

Make this formulation triple-strength when examining 100-ml samples.

The medium is prepared in the following stages:

(a) Dissolve the dehydrated lactose broth and lauryl tryptose broth sequentially in water, without heating.

(b) Dissolve the bromocresol purple in 10 ml of sodium hydroxide solution (4 g of NaOH in 1 litre of water). Sodium hydroxide pellets are caustic and great care should be taken during the preparation of the solution; in particular, gloves and eye protection should be worn.

(c) Add the bromocresol purple solution to the broth solution.

(d) Dispense 50 ml of the medium into screw-cap glass dilution bottles of capacity 250–300 ml. A fermentation tube is not necessary.

(e) Autoclave for 12 minutes at 121 °C, limiting the total time in the autoclave to 30 minutes or less.
(f) Measure the pH of the medium after autoclaving; it should be 6.8 ± 0.2.

## A8.2 Procedure

(a) Mix the sample thoroughly by inverting the sample bottle several times.
(b) Add 100 ml of the sample to the dilution bottle.
(c) Incubate at 35 ± 0.05 °C and examine after 24 and 48 hours.
(d) A positive result (acid production) is indicated by a distinct yellow colour in the medium. Shake the bottle gently and examine for foaming, which indicates the production of gas. Any test in which gas and/or acid is produced should be regarded as a positive presumptive test.
(e) Positive presumptive tests should be confirmed by inoculating a tube of brilliant-green lactose–bile (BGLB) broth with cultures that show acid and/or gas production and incubating at 35 ± 0.5 °C. Growth and the production of gas in the BGLB broth culture within 48 hours confirm the presence of coliform bacteria.

Other indicator bacteria can be detected by the presence–absence test by selecting the appropriate confirmatory medium.

Annex 9

# Residual free chlorine test

The method recommended for the determination of chlorine residual in drinking-water employs *N,N*-diethyl-*p*-phenylenediamine, more commonly referred to as DPD. Methods employing orthotolidine and starch–potassium iodide were formerly also recommended. The first of these reagents is a recognized carcinogen and the method is not reliable. The method based on the use of starch–potassium iodide is not specific for free chlorine, but measures directly the total of free and combined chlorine; it is not recommended except in countries where DPD cannot be obtained or prepared. In this Annex, therefore, only the DPD method is considered.

In the laboratory, photocolorimetry or spectrophotometry may both be used for the determination of chlorine by means of DPD. However, it is common practice and highly recommended for field measurements using simple colour-match comparators to be done on site. The colour is generated following the addition of DPD to the water sample and is matched against standard coloured discs or tubes. The method can be used by staff without extensive specialized training. The reagent may be solid (e.g. individually wrapped tablets) or in the form of a solution; the former is more stable. If the solution is used, it should be stored in a brown bottle and discarded as soon as it starts to become discoloured.

## A9.1 Commercial visual comparator technique

### A9.1.1 Equipment

Commercial comparators are of two basic types—the disc type, containing a wheel of small coloured glasses, and the slide type, containing liquid standards in glass ampoules. However, both consist of the same components: a box with an eye-piece in front and two cells, the whole arranged so that both cells are in the field of vision of the eye-piece.

One cell, containing a water sample without the reagents, is placed in line with the rotating coloured glasses or the ampoules containing the standards. The water sample containing the reagent is placed in another cell. If free chlorine is present, a colour will develop. The concentration of chlorine is estimated by matching the colours in both cells, as seen through the eye-piece. Each colour of the disc or ampoule corresponds to a certain quantity of chlorine in the water; different calibration discs or ampoules are needed for each of the reagents specified.

## A9.1.2 Reagents

Most comparators are intended for use with the manufacturer's own reagents, and care must therefore be taken to keep a good stock of these. This is a disadvantage, since it involves dependence on the local supplier, and importation problems may occasionally arise. On the other hand, it is not necessary to prepare solutions of standards, which makes the technique very easy to use.

## A9.1.3 Method

**A.** Rinse a comparator cell two or three times, and then fill it up to the mark with the water sample.

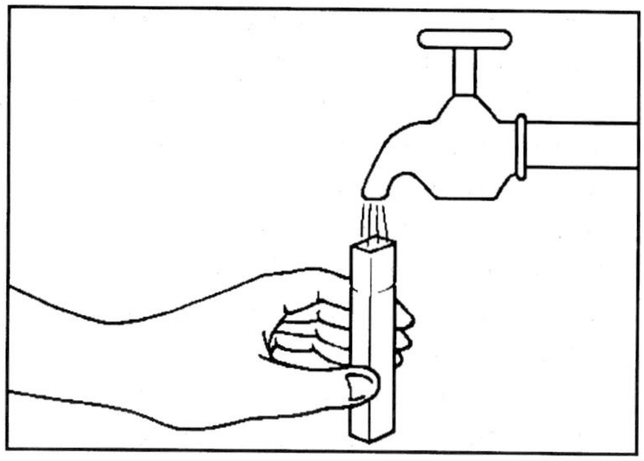

WHO 96645

**B.** Place the cell in the cell carrier of the comparator, which is in line with the coloured standards (B).

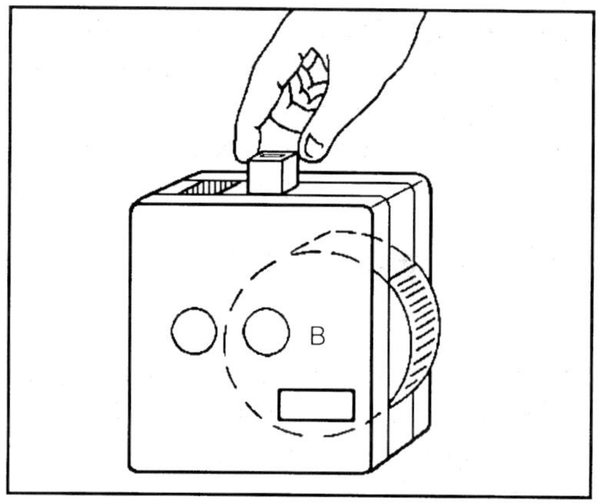


WHO 96646

C. Rinse the second cell and fill it with the same water.

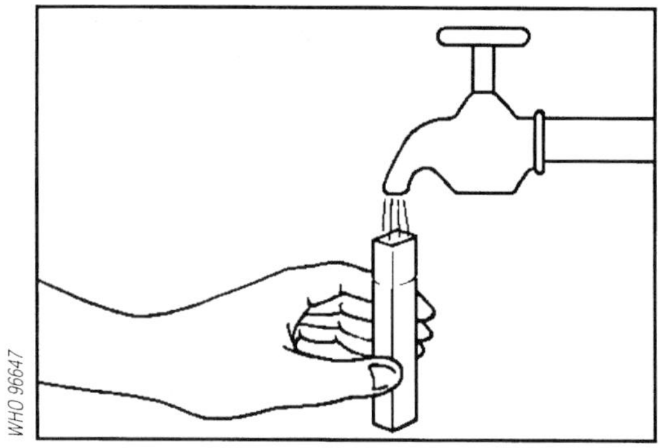


D. Add reagent to the second cell, in accordance with the manufacturer's instructions.

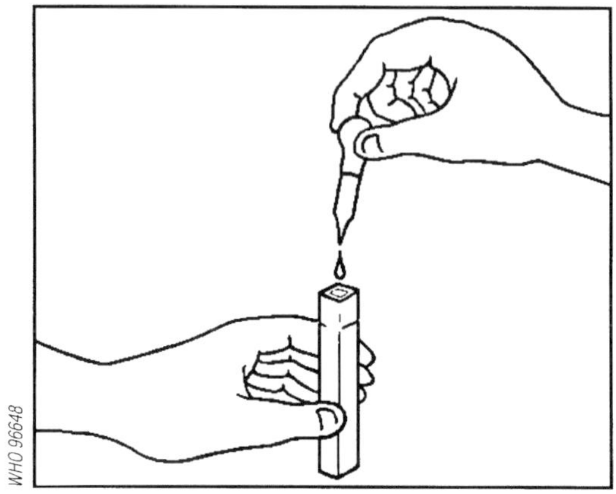


E. Shake the cell (for not more than 3–5 seconds) to mix the reagent.

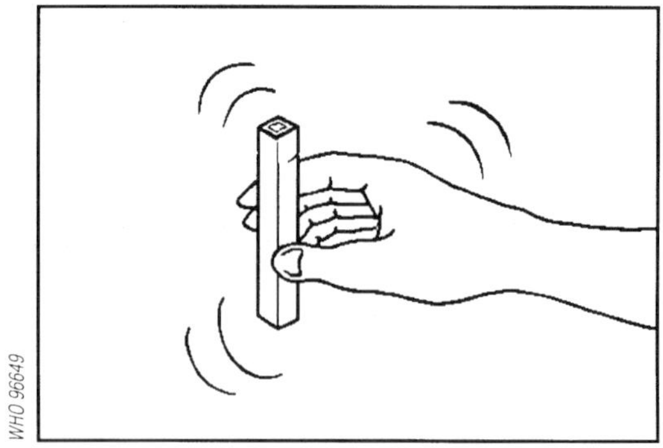

F. Place the cell in the comparator (A).

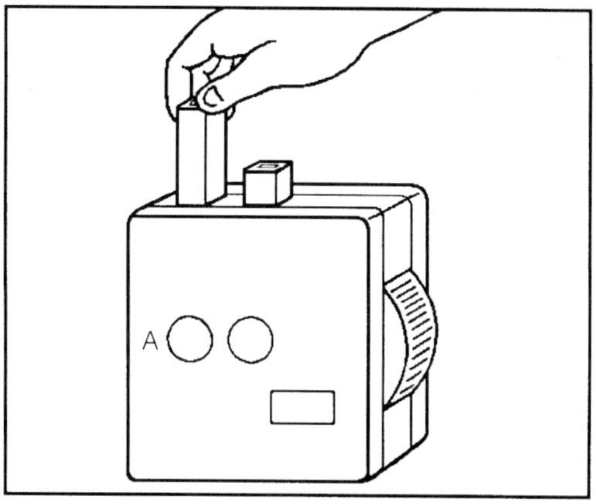


WHO 96650

G. While holding the comparator facing good natural light, rotate the disc until the colour of a standard (B) is the same as that developed by the reagent (A). Immediately (i.e. in less than 20 seconds) read at C the value of free chlorine in mg/litre.

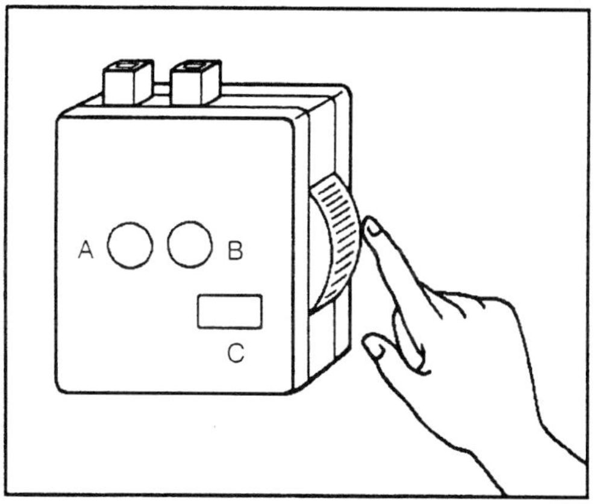


WHO 96651

## A9.2 Colour match comparator method

The procedure employed when a colour-match comparator is used is summarized below. Some comparators employ tubes or discs with the standard colours; the procedure is similar in all cases.

**A.** Rinse the comparator thoroughly in the water to be tested and then fill to the specified lines on the test and control tubes.

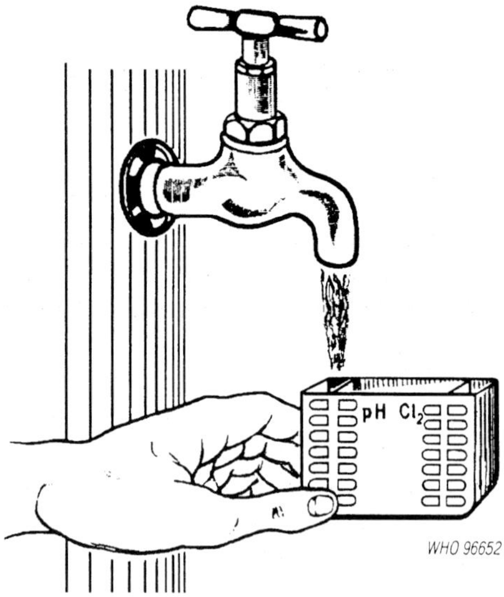


**B.** Add tablet or liquid reagent and mix thoroughly to dissolve. This may require the crushing of the tablet with a clean glass rod.

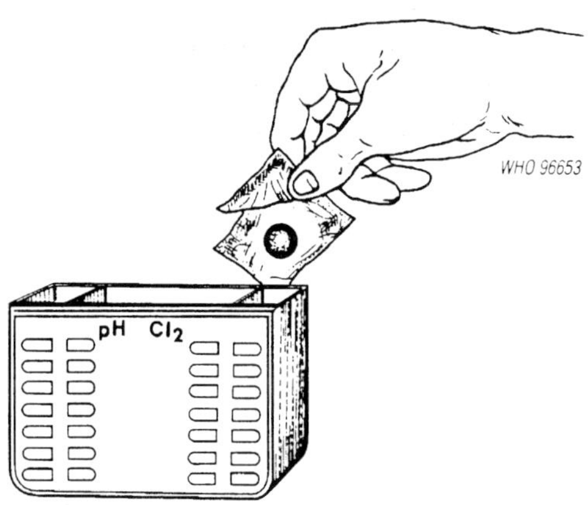


**C.** Compare the pink colour in the test compartment with the standards in the control compartment by viewing the comparator in good, transmitted natural light. Express the result as mg/litre of free residual chlorine.

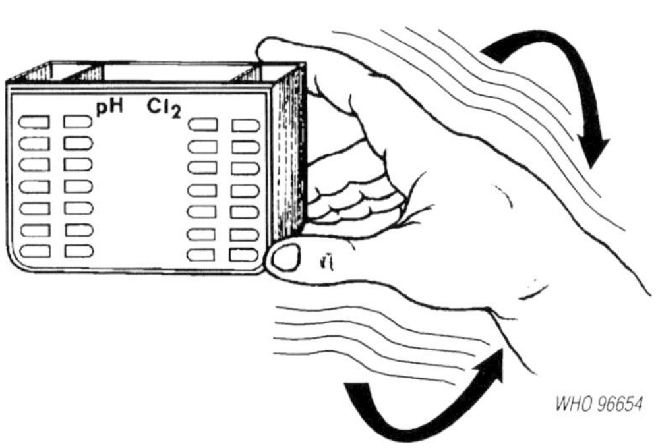

Annex 10

# Turbidity and pH

## A10.1 Measurement of turbidity

High levels of turbidity can protect microorganisms from the effects of disinfection, stimulate the growth of bacteria, and exert a significant chlorine demand. Where disinfection is practised, the turbidity must always be low, e.g. below 5 NTU/JTU, and ideally below 1 NTU for effective disinfection. Measurement of turbidities lower than 5 NTU will generally require electronic meters. However, turbidities of 5 NTU upwards can be measured by simple extinction methods, which are far cheaper and require no consumables. In the monitoring of small-community supplies in developing countries, such methods may be preferable. The sequence of steps involved in turbidity determination by an extinction method is shown below.

A. Add water slowly to the turbidity tube, taking care not to form bubbles. Fill until the mark at the bottom of the tube just disappears.

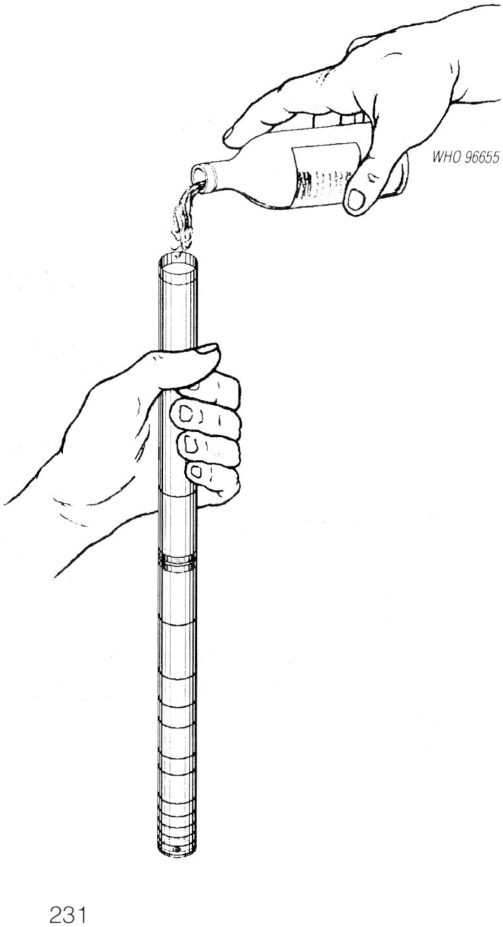

**B.** Read the turbidity from the scale marked on the side of the tube. The value is that corresponding to the line nearest to the level of the water in the tube. The scale is not linear, and extrapolation of values between the lines is therefore not recommended.

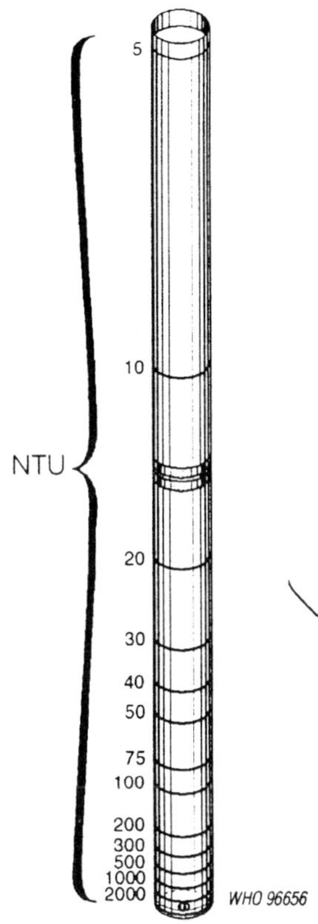


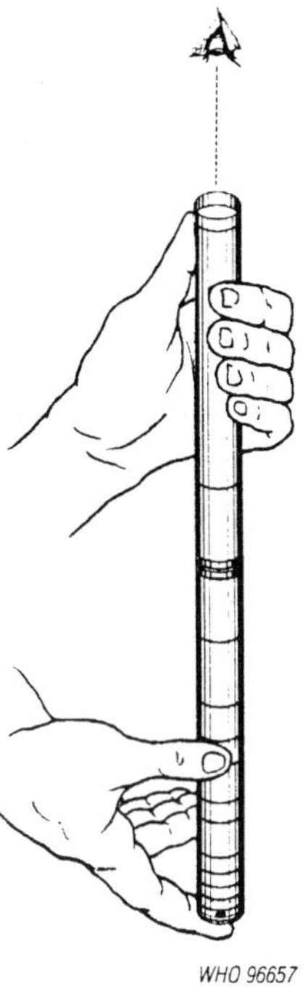


## A10.2 Measurement of pH

### A10.2.1 Electronic pH method

The electronic method of measuring pH requires an electronic pH instrument and electrode, and pH buffer solutions at pH 4.0, 7.0, and 9.0.

A wide variety of pH instruments is available; the less expensive tend to be "disposable" and have a life span of approximately 1 year when used in the field. The more expensive portable models generally have replaceable electrodes, and some may have rechargeable batteries to save recurrent costs.

The most common cause of failure of a pH meter is a damaged electrode; this is generally due to poor storage and maintenance of the electrode when it is not in use. The electrode **must not** be allowed to dry out and **must** be stored in pH 4.0 buffer solution. It must also be protected from impact and vibrations that could crack the glass bulb.

The method of calibration is as follows:

(a) Switch on the pH meter and select pH (if the meter has several functions).
(b) Make sure that the electrode is connected.

(c) Using ready-prepared pH buffer solutions (pH buffer powder mixed with distilled water according to the manufacturer's instructions), place the pH electrode in a pH 7.0 buffer and adjust the meter if necessary.
(d) Rinse the electrode in distilled water and transfer it to pH 4.0 buffer; adjust the meter if necessary.
(e) Rinse the electrode in pH 9.0 buffer and adjust the meter if necessary.
(f) Check the meter in all three buffer solutions. If it does not read true, repeat the above process. If it cannot be adjusted to read correctly in all buffers, suspect a faulty or damaged electrode.

The meter is now ready for use in testing the water sample; calibration of the meter must be carried out daily.

### A10.2.2 Comparator disc method

The comparator disc method for measuring pH requires a comparator, colour discs—depending on the range required (see below)—and the following reagents:

| | |
|---|---|
| universal | pH 4–11 |
| phenol red | pH 6.8–8.4 |
| bromothymol blue | pH 6.0–7.6 |
| bromothymol purple | pH 5.2–6.8 |
| thymol blue | pH 8.0–9.6 |

For most natural waters; the universal reagent and phenol red will be sufficient. Where greater accuracy in a particular range is required, the appropriate disc and reagents should be purchased.

The comparator unit is generally suitable for all the discs and so only one such unit is required. The method of use is similar for all pH ranges:

(a) Place a water sample in the glass or plastic cuvettes provided.
(b) Add the reagent tablets, powders, or drops according to the manufacturer's instructions.
(c) Select the appropriate colour disc and place it in the comparator unit.
(d) Place the cuvettes in the comparator unit.
(e) Hold the comparator unit up to the eye, facing good daylight (but not direct sunlight).
(f) Rotate the disc and observe until the colour matches that of the water sample.
(g) Read the pH value from the disc.

If the pH is not within the range of the disc, select the appropriate reagents and disc and repeat the above procedure.

Annex 11

# Examples of regional and national monitoring report forms for water supplies and for coverage with basic sanitary facilities

This annex contains examples of report forms for a national rural water-supply component inventory (Fig. A11.1), sanitary inspections of gravity-fed supply systems from protected spring sources without treatment (Fig. A11.2), surveillance of rural water-supply quality (Fig. A11.3), and regional and national rural coverage with sanitary facilities (Figs A11.4 and A11.5).

***Fig. A11.1 National rural water-supply component inventory***

| Component | National totals |
|---|---|
| Number of systems | |
| Number of protected springs | |
| Number of surface-water intakes | |
| Treatment plants: | |
| — number of sedimenters | |
| — number of systems with coagulant dosing | |
| — number of systems with a flocculator | |
| — number of systems with slow sand filtration | |
| — number of slow sand filters | |
| — number of systems with rapid sand filtration | |
| — number of rapid sand filters | |
| — number of storage tanks | |

### *Fig. A11.2 Sanitary inspections of gravity-fed supply systems from protected spring sources without treatment*

| Inspection | National totals |
|---|---|
| *Springs:* | |
| — with protection | |
| — with sanitary lid | |
| — locked | |
| — with fence or wall | |
| — with surface-water diversion ditch | |
| — with excreta disposal nearby | |
| *Conduction lines:* | |
| — with visible leaks | |
| *Reservoirs:* | |
| — with sanitary lid | |
| — locked | |
| *Disinfection:* | |
| — with equipment | |
| — with chlorine stock | |
| — operating when inspected | |
| *Adduction lines:* | |
| — with visible leaks | |
| *Distribution networks:* | |
| — with visible leaks | |
| — with constant pressure | |
| *Mean risk score:* | |

### *Fig. A11.3 Surveillance of rural water-supply quality*

**Department:**
**Province:**

| | Community | System type | Source type | Total population | Quality (mean thermotolerant coliforms per 100 ml) | Monthly cost for domestic use[a] | Continuity[b] (overall %) | Quantity[c] (mean litres/day) |
|---|---|---|---|---|---|---|---|---|
| | | | | | | | | |
| Subtotals/ means | | | | | | | | |

[a] Cost is the tariff paid for domestic connection.
[b] Continuity is the overall percentage of time for which water is supplied.
[c] Quantity data are derived from intake volume and do not necessarily reflect the volume devoted to domestic use.

***Fig. A11.4 Regional rural coverage with sanitary facilities***

**Department:**
**Province:**

| | Community | Total population | Water | | Excreta disposal | | | |
|---|---|---|---|---|---|---|---|---|
| | | | Domestic connection | Public standpost | Private latrine | Communal latrine | Septic tank | Sewerage |
| | | | | | | | | |
| Province subtotals | | | | | | | | |
| Department subtotals | | | | | | | | |

### *Fig. A11.5 National rural coverage with sanitary facilities*

| | National totals |
|---|---|
| *Water:* | |
| — by domestic connection | % |
| — by public standpost | % |
| *Excreta disposal:* | |
| — by private latrine | % |
| — by communal latrine | % |
| — by septic tank | % |
| — by sewerage | % |